HETCH HETCHY

HETCH HETCHY

Undoing a Great American Mistake

KENNETH BROWER

Heyday, Berkeley, California

The publishers are grateful to the Kendeda Fund for a generous grant in support of this project.

© 2013 by Heyday
Text © 2013 by Kenneth Brower

All rights reserved. No portion of this work may be reproduced or transmitted in any form or by any means, electronic or mechanical, including photocopying and recording, or by any information storage or retrieval system, without permission in writing from Heyday.

Library of Congress Cataloging-in-Publication Data

Brower, Kenneth, 1944-
 Hetch Hetchy : undoing a great American mistake / Kenneth Brower.
    pages cm
  ISBN 978-1-59714-228-1 (pbk. : alk. paper)
 1. Water-supply--California--San Francisco Bay Area--History. 2. Hetch Hetchy Reservoir (Calif.) 3. Hetch Hetchy Valley (Calif.) 4. Water resources development--Political aspects--California--San Francisco Bay Area. 5. Dam retirement--Environmental aspects--California--San Francisco Bay Area. 6. Environmentalism--California--San Francisco Bay Area--History. I. Title.
  TD225.S25B76 2013
  627'.80979447--dc23
                            2012048053

Cover photo courtesy of Yosemite NPS Library (see p. 73 for full caption)
Cover Design: Lorraine Rath
Interior Design/Typesetting: Jamison Design/J. Spittler
Image Editor: Karen Sorensen

Orders, inquiries, and correspondence should be addressed to:

 Heyday
 P.O. Box 9145, Berkeley, CA 94709
 (510) 549-3564, Fax (510) 549-1889
 www.heydaybooks.com

Printed in China by Regent

10 9 8 7 6 5 4 3 2 1

CONTENTS

INTRODUCTION p. 1

1 STORM LOAF p. 9

2 CRUCIBLE p. 35

3 AGE OF DAMS p. 81

4 AFTER THE DELUGE p. 93

ABOUT THE AUTHOR p. 118

Hetch Hetchy Valley in 1911, by Matt Ashby Wolfskill. *Courtesy of the Prints and Photographs Division, Library of Congress, LC-USZ62-135261*

A modern view of Hetch Hetchy Reservoir. *Courtesy of Anthony Dunn Photography*

INTRODUCTION

For as long as I can remember, more than sixty years and counting, I have known the name Hetch Hetchy. My father took me as a small boy to this flooded Sierra Nevada valley, but I have no recollection of the trip, so in the beginning there was only the word. The word was irritating. "Hetch Hetchy" had a grating, itchy sound, reduplicative and annoying. Had I been told that *hetch hetchy* was an approximation of a Native word for grass, the tall species with edible seeds that once covered the valley floor—Native wheat, in effect—then I might have liked it more.

Chief Tenaya, leader of the Ahwahneechee, the band of Miwoks inhabiting Yosemite, disputed that *hetch hetchy* means grass. *Hetchy* means tree, he said. Redoubled, it refers to a pair of yellow pines that stood in the meadow at the head of the valley. Tenaya may have been right, but he just as easily could have been pulling our leg. The Mariposa Battalion had recently driven Tenaya's people from their home, Yosemite, the most beautiful valley on Earth; the chief could well have left his usurpers a parting gift of misinformation.

But in the end, my real trouble with "Hetch Hetchy" was not in the sound or the translation, but in the connotations, which had gathered darkly about the name by the time I was seven or eight. Hetch Hetchy in our family carries a sense of bitter loss.

In 1952 my father, David Brower, was hired as the first executive director of the Sierra Club, the organization John Muir founded in 1892. My father shared the founder's preference for unflooded valleys. Like Muir, he was a charismatic and an ecstatic, though somewhat less ecstatic than Muir, and at the Sierra Club he restored a Muir-like style of leadership Throughout most of the second half of the twentieth century, it was commonplace for

David Brower at O'Shaughnessy Dam, May 13, 1954, photo by Philip Hyde.
Courtesy of the author

the press to suggest that David Brower was Muir reincarnate. My father himself never expressed any enthusiasm for this notion. But the lives of the two conservationists did run parallel in remarkable ways. Both were autodidacts. Both were hopelessly in love with the Sierra Nevada. Both loved and nurtured the Sierra Club. Both hated dams.

Muir's greatest triumph, his leading role in the fight to make a national park of Yosemite, was followed immediately by the harbinger of his greatest defeat. No sooner had he won federal status for Yosemite than the City of San Francisco, in search of a water source, proposed damming and flooding Hetch Hetchy Valley, in the northeastern part of the new park.

My father's first triumph, his leading role in the successful fight against a pair of dams that would have flooded Dinosaur National Monument in Utah and Colorado, was followed immediately by his greatest defeat, his failure to stop Glen Canyon Dam on the Colorado.

It did not escape my father, as he fought those Colorado dams in the 1950s, that O'Shaughnessy Dam in Hetch Hetchy and Muir's long struggle against it were foreshadowings of his own tribulations; that the dead water of Hetch Hetchy Reservoir made a potent symbol for what he and his movement were up against, not just on the Colorado and its tributaries, and not just with rivers, but with threatened species and ecosystems and landscapes everywhere on Earth.

At the height of his campaign against the Dinosaur dams, my father drove to Hetch Hetchy with his sixteen-millimeter movie camera. Accompanying him was a colleague, the photographer Philip Hyde, who rode shotgun with his 4x5 view camera. In a single day, May 13, 1955, half of which they spent in Hetch Hetchy Valley and half in Yosemite Valley, Phil Hyde shot the black-and-white photos that would illustrate the Hetch Hetchy chapters of the Sierra Club book *The Battle for Yosemite* while my father shot all the footage for a Sierra Club film he called *Two Yosemites*. This shoot-it-all-in-one-day schedule might seem a daring conceptual approach to documentary art—an Andy Warhol sort of film project—but in fact it was just a concession to the fiscal realities of the nascent environmental movement. My father's budget was tiny. He shot five hundred feet of film that day and used almost every inch of it.

Two Yosemites opens with tinny music and the scene of a Yosemite meadow in springtime, a black oak in full leaf in the foreground, conifers beyond. Yosemite Valley on May 13, 1955, looks just as it does in springtime today.

"This is the story of two Yosemites," my father's narration begins. "One is the Yosemite that everybody knows. It is one of the most beautiful valleys in the world. Handsome cliffs and waterfalls. Trees to frame the vistas. Meadows to look at and to look from, natural beauty underfoot and to walk by. A setting with open space, living space, that a million people see and enjoy every year." *On screen, scenes from Yosemite Valley: the vertical wall of El Capitan, waterfalls, lush meadows, conifers reflected in the Merced River.*

"The other Yosemite was only a little less beautiful than this one, and a few miles to the north, Hetch Hetchy Valley."

The scene shifts abruptly northwest to Hetch Hetchy Dome, a rock shaped remarkably like El Capitan, but on a somewhat smaller scale. Down the eastern shoulder of this monolith plunges Wapama Falls, descending in three steps, like Yosemite Falls of the more famous valley, but more thunderously, for Wapama is the mightiest waterfall in all Yosemite National Park. On this 1955 day it roars with spring runoff. The camera follows the falls down through the explosion of spray and mist at the base of the wall. The torrent flows in several channels through the jumble of talus at the foot of the cliff, then suddenly, shockingly, it crosses into the dead, bleached zone of the "bathtub ring" of the reservoir. The stream tumbles on down, and the camera follows, to where Falls Creek meets the reservoir's surface, unnaturally still and green.

"Wapama Falls was one of Hetch Hetchy's finest, but now the setting is gone. You see, Hetch Hetchy was a remarkable storage vessel, with a fine damsite where the walls crowded together. They said the dam would be easily covered by grasses and vines." *The old hand-cranked Bolex movie camera gives the lie to this. It frames the foot of the reservoir, the dam, and the dam's containing walls of granite to show that nowhere, forty years after construction of the dam, is there a trace of the promised grass or vines or a hint of any other vegetation.*

"John Muir, the Sierra Club, and other conservation groups fought hard against this destructive park invasion. San Francisco argued that without this water, it would wither; it must have this cheap power. There were no good alternatives and the dam would enhance the beauty of the place and make it more accessible. The greatest good for the greatest number. Teeming San Francisco against the few people who had yet visited Yosemite. Not one of the City's claims has proved valid, but forty years ago, San Francisco won. This once beautiful valley, part of a national park, was flooded.

"We took these pictures May 13, 1955. The reservoir was down 180 feet, but 60 feet higher than it had been May 1. In June, it would probably fill, then start right down again. For it is a fluctuating reservoir, as most storage and power reservoirs must be. Its zone of ups and downs, between high water and low, is a region of desolation. Nothing permanent can grow in it. This zone is ugly enough at the dam; at the head of the reservoir, it's worse. There, this much drawdown means two miles of desolation, with a vengeance." *The camera pans the 180-foot-tall bathtub ring and the silt-whitened shoreline rubble. A wind sound comes up on the soundtrack, as dust begins to blow on the screen. The silt of the bathtub ring goes airborne. The wind rises and the dust blows harder.*

"We walked a mile up into it," the narration continues. "Though it was early in the season, sixteen hundred people came into Yosemite Valley that same day. Two came to Hetch Hetchy. And we didn't come for pleasure. Who would? What you see here is what you'll see in most fluctuating reservoirs and what no one should see in a park. Stumps where the basin was cleared, stumps and more stumps, exposed and reexposed, until silt finally buries them. The stream, which was one of the most beautiful in the Sierra, is silted in. Tuolumne Falls is covered. The banks are silted. The flat living space is silted. And as soon as the silt is dry enough, it is on the move, a dustbowl, from the silt that sloughed off the canyon sides when the reservoir was full."

On the screen the dust blows harder, scouring across a bank of pale reservoir silt punctuated by stumps left by the logging. The camera lingers on one dark stump. As the wind speed increases, the stump turns ghostly through the thickening scrim of dust.

"They had circulated touched-up pictures of an artist's conception of a pretty lake, always brimful. San Francisco's mayor wrote, 'The scene will be enhanced by the effect of the lake, reflecting all above it and about it, in itself a great and attractive natural object.' Secretary of Interior James Garfield testified that, 'in weighing the two sides of the question, I thought that it should be resolved in favor of San Francisco, because this use of the valley would not destroy it as one of the most beautiful spots in the West. It would simply change the floor of the valley from a meadow to a beautiful lake.' Congressman Englebright added, 'As it is a lake, it will be one of great beauty. There will be fine fishing in it, and boating, and so on.'" *On screen the wind gusts harder. As it accelerates, more and more of the silt of the bank lifts off to become atmospheric. The dust storm blanches into a full whiteout. The ghostly stump turns ghostlier, and soon it is hard to make out. Its existence becomes debatable, like an afterimage on the retina, and then it is gone.*

Two Yosemites is effective filmmaking. It is propaganda, of course, but the best propaganda, as its practitioners point out, is the truth. My father's claims that the bathtub ring was 180 feet tall on May 13, 1955, and that two weeks earlier it had been 240 feet tall, and that the dropping reservoir generates dustbowls, and that nothing permanent can grow in the zone of fluctuation, are not fabrications. They are true. The fabrications were all by the other side—the City of San Francisco and other advocates for the dam. The falsehoods were the fiction of that artist's conception of a lovely brimful lake, and the claim that the dam would be covered with grasses and vines, and the promise of fine fishing and boating. The "lake" is almost never brimful. The bathtub ring is a permanent feature. With each drawdown, dust storms are inevitable. There are no vines. There is little visitation. Neither boating nor swimming is permitted here.

John Muir died, heartbroken, in 1914, the year after passage of the Raker Act, which permitted San Francisco to dam Hetch Hetchy. Many believe that the loss of Hetch Hetchy hastened his death. My father, for his part, was heartbroken by his failure to stop his own bête noire, Glen Canyon Dam—or he was heart*stung*, anyway. He lived for another half century, almost, but he never quite got over the loss. He blamed himself, convinced that if only he had thought faster on his feet in Washington, if only he had been more courageous as the bill came up for a vote in the House, then he could have killed the entire Colorado River Storage Project, all four dams—Glen Canyon Dam included—not just the two dams he did succeed in stopping. Well into old age, when speaking of the beauty of the canyons lost under the surface of Lake Powell, he would get teary and turn away.

For many years the idea of dam removal, on either river, was not something he wanted to contemplate. Part of his reluctance to rejoin the battle, surely, was shell shock. Glen Canyon and its tributaries, the most beautiful sandstone canyons on Earth, were gone. When one loses a great love, whether mate or landscape, survival depends on letting go, on moving on with life.

But part of it, too, I have no doubt, had to do with the utility of tragedy in a profession like my father's. As tales of irredeemable loss, Hetch Hetchy and Glen Canyon made powerful parables. Both river gorges had gone under, but perhaps not entirely in vain; my father told their stories again and again in his speeches, in his congressional testimony, in the books he published, in the forewords he wrote. The two disasters were useful not just in engaging his various audiences—in firing up his rabble—but also in hardening his own resolve never to relent or compromise, never to let this happen again. "Hetch Hetchy's setting is irretrievably lost to all of us, and to all generations," he declared in *Two Yosemites*, and for a long time he believed this to be true.

His change of heart—a new willingness to commit without reservation to the idea of dam removal—came, as nearly as I can figure, sometime in the 1980s.

In 1985 a river engineer, Dave Wegner, took him on a tour of Glen Canyon Dam and a stretch of canyon downstream. Wegner is now staff director for the U.S. House of Representatives Subcommittee on Water and Power, but at the time he was a scientist working on Glen Canyon Environmental Studies, an interagency program charged with investigating the downstream effects of Glen Canyon Dam.

"Downstream of the dam, the high-water releases in 1983 and 1984 mobilized a lot of sediment in the bed in the river and put it up into the water column," Wegner told me. "As the water receded, a layer of silt was left on the red Navajo sandstone in the Glen Canyon reach. Your father asked if the sediment layer on the rock would be left for a long time. I said, no, it would be gone quickly. To prove the point, I took him over to the rock, got a bucket of water and a rag that was in the boat, and began to wash the wall of rock. As the rock dried in the desert sun, you could clearly see that the brown stain of the sediment was gone."

The next day the two men ventured out on Lake Powell in a Bureau of Reclamation boat. My father this time asked Wegner about the permanence of the bathtub ring on this reservoir side of the dam. (Here, upstream, the pale band was not a simple coating of silt, like the mark left downstream by the high-water releases of the previous two years. This ring was chemical, a patina of calcium carbonate and other minerals deposited by the wicking effect of the water flowing back out of the sandstone as the level of the reservoir dropped.) Wegner answered that the ring was not indelible. Time and rain would wash it off.

To prove this, he motored into the Navajo Canyon arm of the reservoir, wetted a sponge, and with a little elbow grease scrubbed the ring away.

Enough time had passed now—two decades since the Glen Canyon Reservoir filled—for my father's scars to have healed, I believe. He was ready to rejoin the battle over dams, and Dave Wegner's two cliff-washing experiments sealed the deal.

In the late 1980s the Sierra Club, in response to the urgings of my father and others, formed the Hetch Hetchy Restoration Task Force, which later became a separate nonprofit, Restore Hetch Hetchy, dedicated to the decommissioning of O'Shaughnessy Dam and restoration of the valley.

"Give the Hetch Hetchy Yosemite back to its original owners—all of us and all our children and theirs and all the natural things that should be living there forever," my father wrote in 1987, in concluding his foreword to a new Sierra Club edition of Muir's *The Yosemite*. "Require San Francisco to get its water from exactly where it now gets it—from the Tuolumne River just below Moccasin Creek powerhouse. More water will be available there if it is no longer wastefully evaporated from Hetch Hetchy Reservoir. The Tuolumne River can be adequately controlled at what was the best place in the first place—the site where Don Pedro Reservoir was built, too late. It holds half a dozen times more water than Hetch Hetchy does. Let San Francisco get its electricity from where everybody else does—not from a national park.

"If Hetch Hetchy is restored and the world has the opportunity to watch the slow but beautifully inevitable recovery of a once sublime valley, you can smile again, John Muir, wherever you are."

David Brower sounded like his old self again.

Even as he helped launch Restore Hetch Hetchy, my father became a founding board member of the Glen Canyon Institute, an outfit dedicated to decommissioning Glen Canyon Dam. His career came full circle: Hetch Hetchy and Glen Canyon became the principal causes of the last months of his life.

The rancorous debate over Hetch Hetchy has had extraordinary persistence. It began in the 1890s, continued throughout the twentieth century, flaring up periodically, fading, then flaring again. Today as I write, thirteen years into the twenty-first century, it is heating up once more.

This is not so hard to understand. There is no more elemental need than our need for water, and San Francisco gets its supply from Hetch Hetchy. Water! Water makes up 60 to 80 percent of the human body. It mediates our baptisms and ablutions. It causes the deserts to bloom. It is the first thing our extraterrestrial probes search for in distant moons and comets as we dowse the universe for other forms of life. Yet Man does not live by hydration alone. In the human imagination the free-flowing river, as source and symbol—as metaphor not just for life, but for existence itself—is a mighty thing, too.

"All the rivers run into the sea, yet the sea is not full; unto the place from whence the rivers come, thither they return again," says Ecclesiastes.

"Life is a river always flowing," says Buddha. "Do not hold onto things. Work hard."

"What is of all things most yielding, can overcome that which is most hard?" asks Lao Tzu. The answer is water. And not just any water, but *running* water.

"And afterwards we would watch the lonesomeness of the river, and kind of lazy along, and by and by lazy off to sleep," says Huckleberry Finn. "Sometimes we'd have that whole river all to ourselves for the longest time. Yonder was the banks and the islands, across the water; and maybe a spark—which was a candle in a cabin window; and sometimes on the water you could see a spark or two—on a raft or a scow, you know; and maybe you could hear a fiddle or a song coming over from one of those crafts. It's lovely to live on a raft. We had the sky up there, all speckled with stars, and we used to lay on our backs and look up at them, and discuss about whether they was made or only just happened."

"Water gives way to obstacles with deceptive humility, for no power can prevent it following its destined course to the sea," says Lao Tzu. "Water conquers by yielding; it never attacks but always wins the last battle."

"Something there is that doesn't love a wall, that wants it down," wrote Robert Frost. Something also there is that does not love a dam.

STORM LOAF 1

The ramble that brought Hetch Hetchy world attention—or to the attention, at least, of that section of North America between the Mississippi and the Atlantic—was John Muir's first walk to the valley, which he wrote up originally in the *Boston Weekly Transcript* of March 25, 1873. That article, "The Hetch Hetchy Valley," entertained Boston readers only briefly over morning coffee before they flipped to the next item, but in later writings Muir would revisit this trip and the subject of the valley again and again.

Setting out for Hetch Hetchy in early November 1871, Muir did not go the way you or I would go today, or even that any normal person would have gone back then—by "the proper route," as Muir himself described it—which would have been the Big Oak Flat Road as far as Hardin's Mills, and then a trail through rock and chaparral to Hetch Hetchy. "As I never follow trails when I may walk the living granite," he informed Bostonians, "I set out straight across the mountains leaving Yosemite by Indian cañon."

Indian Canyon is a steep ravine in the north wall of Yosemite Valley, between Yosemite Falls and Royal Arches. The first stage of Muir's trip, then, was strenuous, a nearly vertical liftoff from the valley floor.

In 1851, twenty years before, Chief Tenaya was brought as a captive to Indian Canyon by Captain Boling, commander of the second military expedition against Tenaya's people, the Ahwahneechee. The remnants of the chief's band were thought to have found refuge in the tableland of low granite domes above the Yosemite Valley rim, and Indian Canyon appeared to be the best way to the top And indeed it was: the Indians had been traveling it since time immemorial. Tenaya, who misled his captors at every opportunity, warned them that Indian Canyon was dangerous and impassible. The climb proved too much, indeed, for Captain Boling, who had to turn back. For the wiry Muir twenty years later, Indian Canyon was easy. On reaching the valley rim, Muir bore off to the left, crossed Yosemite Creek about a mile back from the lip of its 2,425-foot plunge as Yosemite Falls to the valley floor. Then he angled up the side of El Capitan toward the gap through which the Mono Trail passed. It was sundown when he reached the summit. Gathering a bed of fir boughs, he camped.

Muir's November departure was late in the season, with no small risk of snow, but as he was on foot, with no companion to worry about, he went joyous and carefree. He carried two woolen blankets and three loaves of bread. The first two loaves would suffice for the ten-day trip in normal circumstances. The third loaf, a big round one, was what he called his "storm loaf." Snowed in, he could live three extra days on it, six days in a pinch. He carried coffee and a trace amount of protein—a two-ounce mug of Fray Bentos Extracum Carnis of Baron Liebig. Baron Liebig's invention, a concentrated essence of beef, was the Victorian version of freeze-dry. "Thus grandly allowanced," Muir wrote, "I was ready to enjoy my ten days' journey of any kind of calm or storm."

On the morning of the third day, Muir crossed a few more glaciated valleys and came to the steep rim of the Tuolumne Canyon. Scanning the wall, he spotted a likely route to the bottom, descended it a few hundred yards, and struck a well-worn trail winding down just where he wanted to go. "At first I took it to be an Indian trail, but after following it a short distance, I discovered certain hieroglyphics which suggested the possibility of its belonging to the bears."

Whether "certain hieroglyphics" included several of those dark, redolent symbols packed with half-digested manzanita berries and steaming still in the November air, Muir does not say, but he does mention paw prints. ("It was plain that a broadfooted mother and a family of cubs had been the last to pass over it.") The California grizzly, not yet extinct, would wander the Sierra and its foothills for another fifty years until a final Tulare County gunshot sent it on to eternity on the state flag. As any Alaska wilderness hiker knows, a grizzly's broad footprint—claw tips inscribing an arc of pinpricks a full six inches ahead of the pad—has an electric effect on the imagination entirely different from the low-voltage jolt delivered by the track of the black bear. And so it was for Muir.

"A little below this discovery of paws, I was startled by a noise close in front. Of course in so grizzly a place, the noise was speedily clothed upon by a bear skin, but it was only the bounding of a frightened deer which I had cornered, and compelled to make a desperate leap in order to pass me. In its hurried flight up the mountain, it started several heavy boulders, that came crashing and thumping uncomfortably near."

Muir tried several false routes toward the canyon bottom, following seams in the rock until they ended abruptly in precipices that forced him to retreat upward and try again. At midday he found a basin in the rock that held a few cups of water. He boiled some coffee, rested, tried again and again, and finally, on encountering the bear trail once more, he followed it down briskly. The tracks of the mother grizzly and her cubs kept faithfully to the path, "scraping it clear of sticks and pine needles, at steep places, where they had been compelled to adopt a shuffling gait to keep from rolling head over heels. Thin crumbs of dirt, around the edges of their tracks, were still moist."

On reaching the bottom, Muir discovered why all bear trails led down into Tuolumne Canyon. Ground-hugging orchards of manzanita grew on the canyon floor, and fine tall groves of black oaks paved brown underneath with fallen acorns. The drifts of acorns reassured Muir. Grizzlies full of acorns, he reasoned, would have less incentive to eat a human being. He camped by the river in a dense stand of incense cedar, making a bed of the boughs, with "spicy plumes" of cedar as a pillow. Muir was, as usual, simultaneously weary and exalted—his default mode. His fire illuminated the brown columns of the cedars around him, and stars glinted through gaps in the canopy above. The sound of distant cascades on the Tuolumne seemed to him like scraps of song. The soothing hush of the river made him drowsy, and the "breath" of his incense-cedar pillow—for Muir everything in the universe had spirit, was alive—quickly put him to sleep.

"Next morning I was up betimes, ate my usual crust, and stared down the river bank to Hetch Hetchy, which I reached in about an hour. Hetch Hetchy bears are early risers, for they had been out in the open valley printing the hoar frost before I arrived."

Leaving his own prints in the hoar frost of the meadow, Muir was struck, as all visitors after him would be, by Hetch Hetchy's uncanny resemblance to Yosemite Valley. Both valleys, he noted, run generally east to west, with a northward bend in the middle. At Hetch Hetchy's western end stands a formation resembling the Cathedral Rocks of Yosemite, in the same relative spot. On the north side, right where El Capitan would be in Yosemite, towers a similar granite dome, spalled away on the valley side into a similar vertical face. Hetch Hetchy Dome. Off the eastern shoulder of this monolith tumbles Hetch Hetchy Fall, as Muir called it then, Wapama Falls as we call it today, about 1,700 feet high by Muir's estimate. It pummels the valley floor in approximately the spot that Yosemite Falls pummels its own valley.

The Hetch Hetchy floor was apportioned to groves and meadows much as Yosemite's was, with species almost identical: black oaks, live oaks, Douglas fir, scattered sugar pines, and some white fir on the talus slopes and side canyons. Thickets of azalea. Tall-grass meadows edged with tracts of bracken fern, several of which Muir measured at more than eight feet tall. The Tuolumne flowed beneath riverside stands of alder, willow, poplar, and dogwood. The stream had enough gradient in a few places to generate small riffles, but for the most part it ran slowly, "often with a lingering expression, as if half inclined to become a lake." All of which describes exactly how the Merced River flows through Yosemite Valley, no need to change a word.

Muir did notice a few differences. On the northwest side of the valley, which receives the most sun, grew the gray pine, *Pinus sabiniana,* a tree of the hot foothills that does not occur in Yosemite Valley. And the outlet was different. "At the end of the valley the river enters a narrow cañon which cannot devour spring floods sufficiently fast to prevent the lower half of the valley from becoming a lake."

This would prove a fateful disparity, though Muir could have had no inkling of this at the time.

"At present," he wrote, "there are a couple of shepherds' cabins and a group of Indian huts in the valley, which I believe is all that will come under the head of improvements."

This line, too, is full of dramatic irony now.

The year 1873, when Muir's *Boston Weekly Transcript* article broke the news on Hetch Hetchy, was a busy one for the valley. Muir followed that initial East Coast story with a similar account in a Western magazine, *Overland Monthly,* and that same year he returned to Hetch Hetchy with a friend, the painter William Keith, a fellow Scot with whom he made many Sierra trips. Also in 1873, as it happened, the painter Albert Bierstadt, famous for his Yosemite Valley landscapes, made a six-week pack trip to the Hetch Hetchy country with three friends, his wife, and his sketchbooks. Upon returning to civilization, he made at least four paintings of Hetch Hetchy. There were now pictures to go with Muir's words.

Often it seems that in some subliminal way Muir anticipated the trouble ahead for Hetch Hetchy—that his recruitment of painters and his introduction of other citizens to the valley arose from an intuition that it would someday need a constituency. A reader senses, even in Muir's earliest Hetch Hetchy essays, that he was already lobbying for the place, polishing his argument.

"It is estimated that about 7000 persons have seen Yosemite," he wrote in that first piece for the *Boston Weekly Transcript*. "If this multitude were to be gathered again, and set down in Hetch Hetchy, perhaps less than one percent of the whole number would doubt their being in Yosemite. They would see rocks and waterfalls, meadows and groves, of Yosemite size and kind, and grouped in Yosemite style. Amid so vast an assemblage of sublime mountain forms, only the more calm and careful observers would be able to fix upon special differences."

Muir was polishing his argument a little too bright. It is undeniable that Hetch Hetchy was an extraordinarily beautiful valley—lift your eyes today from the reservoir and bathtub ring and you can see what a magnificent place it must have been—but Hetch Hetchy is also unique, and the ice carved it on a different scale from the more celebrated valley to the southeast. Any observer fully awake and sober knows right away that she or he is not in Yosemite and quickly fixes on those "special differences."

"Almost an exact counterpart of the Yosemite," wrote Josiah Whitney, chief of the California Geological Survey, in one of the first published accounts of Hetch Hetchy. "It is not on quite as grand a scale as that Valley; but if there was no Yosemite, the Hetch Hetchy would be fairly entitled to a world-wide fame, and, in spite of the superior attractions of the Yosemite, a visit to its counterpart may be recommended, if it be only to see how curiously nature has repeated herself."

"The walls of this valley are not quite so high as those of Yosemite," wrote the painter Alfred Bierstadt, "but still, anywhere else than California, they would be considered as wonderfully grand. It is smaller than the more famous valley, but it presents many of the same features in its scenery and is quite as beautiful."

"It is much smaller than the Yosemite," wrote the Yosemite Commissioner Ben Truman, "and, therefore, many of its objects are grouped together very grandly and very beautifully, and at once entrance the beholder; but Hetch Hetchy lacks many of the imposing features of the Yosemite. Still, if there had been no Yosemite, Hetch Hetchy would command the admiration of all who visit it, and would probably rank as the grandest and most beautiful aggregation of rock and water in the world—in fact, it would be Yosemite."

If most observers were somewhat more measured than Muir in their claims for Hetch Hetchy, at least a couple were more rabid.

"Volumes have been written descriptive of Yosemite," wrote a *San Francisco Chronicle* correspondent, W.P.B. "But of her sister valley, equally beautiful, and in some respects even more remarkable, our public prints, our books of travel, and with one exception, even our guide-books are silent. Here is a valley, which, were it in Europe, Americans would cross the ocean by thousands to see. Yet, lying as it does, at our very doors, we doubt if one Californian in a hundred knows of its existence. From a rocky bluff, a half-mile down the cañon, Hetch Hetchy comes first into open view. It is a surprise. The panorama is a noble one, embracing in one vast amphitheatre all the most notable objects of interest of the Valley. Yosemite cannot produce its equal."

"Hetch Hetchy is superior to Yosemite," one travel writer, Xenos Clark, flatly declared. "Yosemite is a long, strung-out cluster, too rambling and too extensive for a single sweep of the eye; moreover, the landscape-gardening of Yosemite is very rude, it is more like an area of enclosed country with its forests and its rough places, traversed by the Merced River. Hetch Hetchy, on the other hand, makes a picture."

Yosemite does not make a picture? Xenos Clark would have had an argument from a legion of great landscape artists—Alfred Bierstadt, William Keith, Thomas Hill, Thomas Ayres, Chiura Obata, and Ansel Adams among them—but he was entitled to his opinion, of course.

One day John Muir, visiting the studio of the artist William Keith, paused before a Hetch Hetchy painting in progress on the easel and complained that Wapama Falls was too slender. Anyone familiar with the great depths of snow that pile up in this watershed, claimed Muir, would know that Wapama must channel a thundering cataract, not this wispy filament. Keith defended his work with mock indignation, but in the end grabbed his brush and thickened Wapama, adding a few thousand cubic feet per second to the falls.

This is what Muir habitually did in his writing on the valley: add a few thousand cubic feet per second. He liked to go a wee bit over the top.

In his book *The Yosemite*, Muir, reworking his Hetch Hetchy material one more time, wrote an account of Tueeulala, the less voluminous but taller waterfall just west of Wapama, describing it as a natural wonder comparable only to Bridal Veil Falls in Yosemite Valley, but excelling "even that favorite fall both in height and airy-fairy beauty and behavior."

The phrase "airy-fairy behavior" is probably not one we would use today—not in connection with a waterfall, anyway—but in fact it captures nicely the whimsy of the Sierra's taller and more wraith-like falls on a day of wind.

"Lowlanders," Muir went on, "are apt to suppose that mountain streams in their wild career over cliffs lose control of themselves and tumble in a noisy chaos of mist and spray. On the contrary, on no part of their travels are they more harmonious and self-controlled. Imagine yourself in Hetch Hetchy on a sunny day in June, standing waist-deep in grass and flowers (as I have often stood), while the great pines sway dreamily with scarcely perceptible motion. Looking northward across the Valley you see a plain, gray granite cliff rising abruptly out of the gardens and groves to a height of 1800 feet, and in front of it Tueeulala's silvery scarf burning with irised sun-fire. In the first white outburst at the head there is abundance of visible energy, but it is speedily hushed and concealed in divine repose, and its tranquil progress to the base of the cliff is like that of a downy feather in a still room."

This is classic Muir. "Silvery scarf burning with irised sun-fire" is prose more purple than we like it today, yet this passage is built on close observation of very tall Sierra waterfalls and conveys a fine sense of their dynamics and rhythm. It allows readers a shock of recognition at how much of this we have noticed ourselves—that ponderous weightlessness of the falling water, that slow-motion drift toward earth.

It is curious how often Muir's Hetch Hetchy writing veered toward apotheosis. What was it that compelled him so often to gild the Mariposa lily of this valley? He was one of those figures in whom the poet and wilderness advocate are melded. Thoreau. John Burroughs. Aldo Leopold. With these figures it is hard to know which came first, the poetry or the advocacy, and it is difficult to tell which tendency at a given moment is in ascendancy. I know, having been raised by one of these people myself.

But in the end it is probably foolish to read any sort of premonition into Muir's brief for Hetch Hetchy. He wrote about many other places in the Sierra Nevada with equal fervor. He was an advocate for the entire cordillera he called "the Range of Light." In all his writing Muir displayed two modes: ecstatic praise of nature alternated with calmer, more dispassionate reporting on the particulars of natural history. It is from this one-man, two-part harmony of Muir's, more than from anything contributed by visiting painters, photographers, and scientists, that we know what Hetch Hetchy was like before the dam.

"The floor of the Valley is about three and a half miles long, and from a fourth to half a mile wide," he wrote. "The lower portion is mostly a level meadow about a mile long, with the trees restricted to the sides and the river banks, and partially separated from the main, upper, forested portion by a low bar of glacier-polished granite across which the river breaks in rapids.

"The principal trees are the yellow and sugar pines, digger pine, incense cedar, Douglas spruce, silver fir, the California and golden-cup oaks, balsam cottonwood, Nuttall's flowering dogwood, alder, maple, laurel, tumion, etc. The most abundant and influential are the great yellow or silver pines like those of Yosemite, the tallest over two hundred feet in height, and the oaks assembled in magnificent groves with massive rugged trunks four to six feet in diameter, and broad, shady, wide-spreading heads. The shrubs forming conspicuous flowery clumps and tangles are manzanita, azalea, spiraea, brier-rose, several species of ceanothus, calycanthus, philadelphus, wild cherry, etc.; with abundance of showy and fragrant herbaceous plants growing about them or out in the open in beds by themselves—lilies, Mariposa tulips, brodiaeas, orchids, iris, spraguea, draperia, collomia, collinsia, castilleja, nemophila, larkspur, columbine, goldenrods, sunflowers, mints of many species, honeysuckle."

This, then, in the words of its foremost advocate, was Hetch Hetchy before the dam.

From a 1909 photograph album. *Courtesy of The Bancroft Library, University of California, Berkeley, BANC PIC 1971.031, 1909.08*

Albert Bierstadt (American, b. Germany, 1830–1902). *The Hetch-Hetchy Valley, California.* c. 1874–1880. Oil on canvas. 37 5/16 x 58 5/16 inches. Bequest of Laura M. Lyman, in memory of her husband, Theodore Lyman. *Courtesy of Wadsworth Atheneum Museum of Art / Art Resource, NY*

Albert Bierstadt (American, b. Germany, 1830–1902). *Hetch Hetchy Canyon*. 1875. Oil on canvas. Gift of Mrs. E. H. Sawyer and Mrs. A. L. Williston. *Courtesy of Mount Holyoke College Art Museum, South Hadley, Massachusetts. Photograph Laura Shea 1876.2.I(b).PI*

A pre-dam view of Hetch Hetchy Valley, c. 1900. *Courtesy of the Yosemite NPS Library*

Sketch by Laura Cunningham

"WHEN THE FIRST PEOPLE LIVED HERE, THEY WERE CALLED THE AHWAHNEECHEE—THE PEOPLE OF THE DEEP GRASSY VALLEY. AND THE NEW PEOPLE CAME AND CALLED THEM YOSEMITE INDIANS. AND THEY ARE CALLED MIWOK AND PAIUTE NOW. THIS WAS ONE OF THE LAST SITES OF THE AHWAHNEECHEE PEOPLE. THEY LIVED HERE AND THEIR SPIRITS ARE STILL HERE. SOMETIMES IF YOU'RE REAL QUIET, YOU CAN HEAR THEM."

—Julia Parker

Sketch by Laura Cunningham

Hetch Hetchy Dome

Wapama Falls was briefly called "Macomb Falls," after Lieutenant Montgomery Macomb, the commander of a survey team that mapped the Yosemite region in the 1870s.

All photographs on this spread courtesy of The Bancroft Library, University of California, Berkeley, BANC PIC 1954.008—PIC Box 3

"The Pack, Hetch Hetchy"

Resting after work
Hetch Hetchy

AMONG THE EARLY NON-NATIVE PEOPLE TO SEE HETCH HETCHY WERE GEOLOGICAL SURVEY TEAMS.

This c. 1908 photograph of Muir next to a large pine tree was likely taken in Hetch Hetchy. *[MSS048f54-3112] John Muir Papers, Holt-Atherton Special Collections, University of the Pacific Library ©1984 Muir-Hanna Trust*

"THE WORLD IS SO RICH AS TO POSSESS AT LEAST TWO YOSEMITES INSTEAD OF ONE," SAID HETCH HETCHY'S GREATEST ADVOCATE, NATURALIST JOHN MUIR, REMARKING ON THE STRIKING SIMILARITIES BETWEEN HETCH HETCHY AND YOSEMITE VALLEYS.

This drawing of Hetch Hetchy Valley waterfalls was created by Muir c. 1877. *[MSS 048 MuirFiche6Frame0307] John Muir Papers, Holt-Atherton Special Collections, University of the Pacific Library ©1984 Muir-Hanna Trust*

Muir with a Sierra Club group on the trail to Hetch Hetchy. *[MSS048.f25-1404] John Muir Papers, Holt-Atherton Special Collections, University of the Pacific Library ©1984 Muir-Hanna Trust*

William Keith (American, 1838–1911). *Hetch Hetchy Side Canyon, I.* ca. 1908. Oil on canvas. 22 x 28 inches. *Courtesy of The Fine Arts Museums of San Francisco, presented to the City and County of San Francisco by Gordon Blanding, 1941.4*

John Muir and William Keith posed with their artistic contemporaries in this early 1900s photograph. From left: Charles Keeler, John Muir, John Burroughs (seated), William Keith, and Francis Fisher Browne. *Courtesy of Yosemite NPS Library*

"Fall Cliffs from between oaks on North side," by J. Le Conte, 1908. *Courtesy of The Bancroft Library, University of California, Berkeley, BANC PIC 1971.031.1908.07*

"Hetch Hetchy from Lower End," from a Sierra Club photograph album, 1909. *Courtesy of The Bancroft Library, University of California, Berkeley, BANC PIC 1971.031.1909.02*

View of Kolana Rock and river, by J. Le Conte, c. early 1900s. These prints from LeConte's negatives were made by Ansel Adams. *Courtesy of The Bancroft Library, University of California, Berkeley, BANC PIC 1971.071*

"Along the river, Hetch Hetchy," by Herbert Gleason, 1909. *Courtesy of The Bancroft Library, University of California, Berkeley, BANC PIC 1971.031. 1909.09*

"Upper meadows, Hetch Hetchy," by Herbert Gleason, 1909. *Courtesy of The Bancroft Library, University of California, Berkeley, BANC PIC 1971.031.1909.09*

"The Ferry Godfather and his children," from a Sierra Club Outing album, 1911. *Courtesy of The Bancroft Library, University of California, Berkeley, BANC PIC 1971.031, 1911.14*

From a Sierra Club Outing album, 1911. *Courtesy of The Bancroft Library, University of California, Berkeley, BANC PIC 1971.031, 1911.14*

"An Afternoon tea party in Hetch Hetchy," from a Sierra Club Outing album, 1909. *Courtesy of The Bancroft Library, University of California, Berkeley, BANC PIC 1971.031, 1909.08*

"Along the banks of Tuolumne River at camp, in Hetch-Hetchy Valley," from a Sierra Club Outing album, 1911. *Courtesy of The Bancroft Library, University of California, Berkeley, BANC PIC 1971.031, 1911.08*

BULLETIN No. 2 OF THE NATIONAL COMMITTEE FOR THE PRESERVATION OF THE YOSEMITE NATIONAL PARK

COMMENTS OF THE UNITED STATES PRESS
ON THE
INVASION OF THE YOSEMITE NATIONAL PARK
AS PROPOSED IN THE HETCH-HETCHY BILL, WHICH HAS PASSED THE HOUSE OF REPRESENTATIVES AND COMES BEFORE THE SENATE DECEMBER 1st TO 6th.

These Editorial Comments Are Entirely Spontaneous Expressions of National Opinion on a Thoroughly Dishonest Bill. They Are Inspired Also by a Strong and Almost Universal Sentiment as to the Danger of Invading Our National Parks.

Enormous Power of Precedent

Boston Post.—The day is coming when, if this impudent attempt to destroy a lovely natural valley, one of the finest in the world, succeeds, there will be set in motion the same sort of machinery that has worked the Hetch-Hetchy grab through the National House, and it will be against other cherished possessions of the people. Few realize the enormous power of precedent with Congress.

The "Beautiful Lake" Delusion

The New York Times.—A prominent advocate of the project has confessed privately that "there are bad things in the bill, but they were put there to get votes." The House debate gives reason for thinking that the measure is a clumsy and probably unworkable attempt to partition the flow of the Hetch-Hetchy watershed between the city and such of the San Joaquin Valley farmers as could thus be bribed to forego their opposition.

The act creating the Yosemite National Park sets forth the importance and duty of reserving these wonders "in their original state," and the world has a moral right to demand that this purpose shall be adhered to. The "beautiful lake" theory deceives nobody. An artificial lake and dam are not a substitute for the unique beauty of the valley.

Project Has a Bad Look

Cleveland Plaindealer.—The whole project has a bad look. Let San Francisco go elsewhere for her water; engineers agree that other sources are available.

In the opinion of the Plaindealer, the best sentiment of the American people, could it be ascertained, would be overwhelmingly against the Hetch-Hetchy proposal.

The name is about all the water grabbers propose to leave ungrabbed.

No Time to Give Away Park Property

Syracuse Post Standard.—We don't believe that the time has come for giving away a national park for any purpose; nor do we admire the spirit which seems to animate San Francisco in this matter.

The Redwoods Will Go Next

Atlanta Morning Journal.—If San Francisco succeeds in stealing the Hetch-Hetchy Valley no doubt she will next want to cut down the redwood trees to obtain timber with which to dam it up.

Irremediable Destruction

Milwaukee Journal.—One fact alone should weigh enough to decide the issue. The cost of another site for waterworks for the city of San Francisco, no matter how great, would some time be paid off. Never will the beauty of the Hetch-Hetchy Valley be regained. With all the natural beauties of a rich nation, one little spot of rare beauty that was the common heritage of all will have been forever blotted out. And for the sake of saving money to a single city.

Almost a Crime

Memphis Appeal.—To any one who has stood in this wonderful valley and has gazed in awe upon the beautiful handiwork of God, it seems almost a crime to consider such a commercial proposition as the one offered by San Francisco.

It has been fully appreciated by sane participants in the general conservation movement that sentiment must not be allowed to run away with common sense, but here is required a wholly unnecessary surrender of a wonderful heritage of Nature, and to grant the request of San Francisco would be a hesitating reflection on the national idealism and an ominous sign.

A Sordid Scheme

Outdoor World and Recreation.—Stripped of specious argument and sentimental enthusiasm, the naked, sordid fact stands revealed that San Francisco seeks to utterly destroy a precious wonderland because it offers cheaper water than can be had elsewhere.

In a word, they want the lovely Hetch-Hetchy Valley together with 500 square miles or half of the Yosemite National Park, which Congress in 1890 dedicated forever to public use!

Too Few Public Parks

Rochester Union-Advertiser.—Not many of those who are interested in seeing this big

A National Scandal

The Standard Union, Brooklyn.—The Hetch-Hetchy matter at Washington has become a national issue and the manner of its handling almost a national scandal. Any citizen, indifferent to either aspect, thinking that it is only of local concern to San Franciscans, or that it is a fair example of parliamentary practice and precedents, greatly errs, and neglects incidents fraught with much significance and portending grave consequences. If the San Francisco "combine" breaks down the guards which the nation has placed around its Yosemite reserves and preserves, all others are in peril, and the whole conservation policy, which has gained its place fighting every step of the advance, goes by the board.

An Outrage and a Crime

Philadelphia Record.—At the hearing before the Public Lands Committee, the most conspicuous advocate of the scheme was asked whether he could not "go out overnight anywhere along the Sierra and get an abundant supply of pure water for the city?" His answer was: "Yes, by paying for it."

The pressing necessity of robbing the American people of a glorious possession evidently does not exist.

The sacrifice of a veritable temple of Nature to commercial greed would not be merely an inexcusable act of folly; it would be an outrage and a crime.

A Crisis

Boston Post.—Washington ought to sit upon this scheme hard. Congress should say once and for all that the great Federal [park] reserves are to be kept intact for the delight of posterity and the admiration of the world.

Irremediable Mischief

New York Evening Post.—The thing proposed in the bill is that the nation shall give up, for the economic advantage of a municipality, one of its most wonderful scenic possessions. Once done, the mischief can never be undone.

An Anti-Conservation Raid

Boston Transcript.—If this measure passes the Senate it will mark the beginning of an anti-conservation raid which has long been planned by those who thirst not for the water, but for its power, and who hunger nightly for the return of the good old days when the resources of the public domain were open to the exploitation of the man who was wily enough to get there first. If Congress surrenders the Hetch-Hetchy it will mean that in its eyes the $42,000,000 worth of water power that it can produce is more valuable than the life-giving refreshment of its unique scenery. It will mean that [...] its self-ie political sophistry has triumphed over the public interest.

An Attempted Steal

Springfield Republican.—The evidence increases that San Francisco's attempted steal of the Hetch-Hetchy Valley is largely motivated by the hope of obtaining not merely its water for ordinary purposes, but its immensely valuable power rights—estimated to be worth $45,000,000—free of cost.

Altogether Reprehensible

Boston Record.—The fight on the mischievous and altogether reprehensible Hetch-Hetchy bill in the Senate is on again today. This bill aims badly and boldly to rob the public, whose possession the Yosemite National Park is, to pander to the greed of San Francisco, which wants a water supply cheap.

Cannot Justify the Spoliation

Brooklyn Daily Eagle.—Only the most urgent necessity could excuse the destruction of Yosemite Park. San Francisco has not proved either an immediate or ultimate necessity for taking water from the Hetch-Hetchy Valley. Competent authorities aver that she can obtain it more quickly and more cheaply elsewhere. But even if this were not so, it is obvious that neither convenience nor economy can justify the spoliation of a national asset to serve a municipal need.

Develop the Parks, Not Destroy Them

Denver Republican.—There is a very strong public feeling in favor of keeping the nation's parks intact, as heritage of pleasure. This feeling will grow stronger with the years and as the public makes a more general use of the parks. The cheapening of transportation, and

CRUCIBLE 2

The twelve-year battle for Hetch Hetchy, in the view of one school of historians, was the protracted labor that marked the birth of the environmental movement. There is no shortage of differing opinions, of course. Good arguments are made for a number of other points along the timeline.

Thoreau's *Walden*, published in 1854, and George Perkins Marsh's *Man and Nature*, published in 1864, were urtexts, certainly, epochal books that introduced the principles that would drive the movement. Later in the nineteenth century, the establishment of Yosemite, Yellowstone, and Sequoia National Parks ceded actual territory, thousands of square miles of it, to those principles. Early in the twentieth, Teddy Roosevelt combined love of the outdoors, eloquence in defense of it, and decisive action to embark on a spree of park creation that has not been equaled since. Some historians point to the successful campaigns against dams on the Colorado Plateau in the 1950s and 1960s as signaling the birth of the movement. Some point to Rachel Carson and the publication of *Silent Spring* in 1962. Some point to the first Earth Day in 1970.

These latter benchmarks, those of the second half of the twentieth century, come too late; to credit them as the starting point is to slight all the groundwork that went before. The argument for the Hetch Hetchy struggle and its period—fifty years after Thoreau and fifty years before the victories on the Colorado Plateau—is compelling. If any one beginning can be singled out, perhaps this is indeed the one. In a remarkable number of ways, the Hetch Hetchy Valley was cradle and crucible The environmental movement as we know it today was forged in the fight against Hetch Hetchy Dam (as it was then called) in the years just prior to the First World War.

Bulletin No. 2 of The National Committee for the Preservation of the Yosemite National Park.
Courtesy of The Bancroft Library, University of California, Berkeley, pff F869.S3.8.C632

No previous debate over the American landscape had so engaged and enraged the American public. The Sierra Club and other opponents of the dam helped stir this sentiment, and then they stoked it. Their campaign to preserve Hetch Hetchy became the prototypic environmentalist campaign. John Muir, William Colby, Joseph N. LeConte, and other luminaries of the Sierra Club reached out to form alliances with similar outfits across the country: the Appalachian Mountain Club, the Society for the Preservation of National Parks, the General Federation of Women's Clubs, the American Civic Association, and others. This coalition launched a multimedia Hetch Hetchy campaign: pamphlets, magazine stories, newspaper editorials, letters to the editor, telegrams to Congress and the president. In 1909 the Sierra Club experimented with another sort of mobilization, diverting the ninth of its annual "High Trips" into Hetch Hetchy so as to familiarize the membership with the valley and build a constituency for the place.

"We are taking the Club Outing into the Yosemite National Park," William Colby, the Club's young recording secretary, explained to an ally. "We'll spend the concluding week in Hetch Hetchy, in order to have the opportunity of getting as many photographs of the place as possible, and educating our members to look at the matter from our view-point."

My own first memories are of the High Sierra and these High Trips. On one of them, in the early 1950s, an elderly Will Colby came along. Colby is my only direct connection with Victorian environmentalism, my sole opportunity to describe, as eyewitness, a protagonist in the opening battle of the long fight for Hetch Hetchy.

The William Colby I knew was foremost among the patriarchs. The third president of the Sierra Club, he had also served as the Club's secretary for forty-six years and would live to eighty-nine. The High Trips were his invention; he organized the first one in 1901. The age spread on these expeditions, one of their great virtues, was accomplished by mules, which carried the kitchen and most of the walkers' gear into the high country, thereby accommodating those who could not backpack much weight, like the seventy-five-year-old Colby and the six-year-old me. Each summer through the 1950s, my father, now the Sierra Club's executive director, escaped the office to lead one or more of these trips. I remember his reverence for Colby and his respect for the old man's advice.

William E. Colby had a fine, strong face immortalized by Ansel Adams, Cedric Wright, and other black-and-white photographers of the Sierra, who could not resist turning their big 4x5 view cameras from the peaks to the man. ("You knew who he was without inquiry," Ansel Adams wrote of Colby. "He comes with him a deep humanity, and the mood of rivers and forests and clean white stone." The imagery was spillover from Adams's darkroom; he might as well have been describing one of his own iconic portraits of El Capitan or Half Dome.) By the time I knew Will Colby, his face was as deeply etched and weathered as a snag of whitebark pine at timberline. The Sierra Nevada is a desert range, shadeless above 11,000 feet, and Colby always wore a bandana neckerchief against the sun. He was slow. When you are six years old, seventy-five seems ancient. It was like hiking with Methuselah in the Sinai. When you looked back to see Will Colby following, several switchbacks below you on the way up to the pass, it seemed that Muir himself could not be far behind.

Back at the turn of the century, the aging Muir and the young Colby had been slow to recognize the gravity of the threat to Hetch Hetchy. The valley had first been proposed as a damsite in 1890, but only as one of a number of possibilities. Muir and Colby were preoccupied in the first months of 1900 by their campaign for "recession" or "retrocession" of Yosemite—reversion of the park from the state to the federal government. As they deliberated at the Sierra Club office in downtown San Francisco, Mayor James D. Phelan and other city officials were themselves deliberating a stone's throw away, quietly laying plans to get the city's water from Yosemite.

In Washington, Mayor Phelan's representatives were busy investigating possible ways to get a legal foothold in the national park. Out West, the mayor's associates had hired the engineer J.B. Lippincott to conduct surveys of reservoir sites on the Tuolumne, with the advice that he work quietly, so as not to tip off the acting superintendent of Yosemite Park or start a rush on Tuolumne water rights. Lippincott was a colleague of William Mulholland, the head of the Los Angeles Department of Water and Power, the man whose dams and aqueducts, and whose devious acquisition of water rights, allowed Los Angeles to grow into a megalopolis. Mulholland diverted the water from the farmland and ranchland of the Owens Valley, on the east side of the Sierra, and sent it south to his city. As Los Angeles bloomed, spreading to the margins of its basin and the limits of the inversion layer that trapped its smog, the Owens Valley withered.

"Forget it, Jake. It's Chinatown," a sympathetic but cynical cop tells Jack Nicholson at the end of the movie *Chinatown,* a fictionalized account of water piracy by the city of Los Angeles. Its villain, "Hollis I. Mulwray," is an imperfect anagram for William Mulholland. "Chinatown," in the L.A. police lingo of the movie, is the code word and metaphor for the way things really get done in Los Angeles. Up in San Francisco, the leaders of the Sierra Club had their own Chinatown, of course, a big one just a few blocks from the office, but they were not yet savvy about that other, figurative Chinatown—the one with origins in realpolitik south of the Mulholland Tunnel.

In May 1900, a congressman from Stockton, Marion DeVries, introduced H.R. 11973, a bill authorizing the secretary of the interior to grant rights of way through government reservations, with specific mention of the three California parks, for "canals, ditches, pipes and pipe lines, flumes, tunnels, or other water conduits, and for domestic, public, or other beneficial uses." Whether or not San Francisco politicians had anything to do with this bill, there is no doubt that H.R. 11973 served the City's secret agenda. It was stealth legislation. The nascent California environmental community, unaware of the legislation, did not act at all.

"We, who were trying to the best of our poor ability to save these great parks for the people, knew nothing of the bill until it was a law," William Colby would write. "We were probably not vigilant enough, but we certainly did not lack the desire to know all that was going on. Be that as it may we later examined the Act and found that it was the intention to still 'preserve and retain' the natural curiosities and wonders of the Park in their natural condition, and that such rights of way as were contemplated should not interfere with 'the attainment of the purposes for which the various reservations are established.' We felt that this apparently harmless act could not injuriously affect the park."

Colby was then just twenty-five. He had graduated from law school in 1898, just two years before, and he had joined the Sierra Club that same year. The organization itself was only eight years old. Colby had been its secretary for just three weeks when Congressman DeVries introduced H.R. 11973. Young Colby was nicely representative of his movement, such as it was—still very green, political instincts unformed.

Shortly after the Right of Way Act passed in February 1901, Colby and Muir and their conservationist colleagues realized that DeVries's little law was not harmless at all. The Hetch Hetchy skirmish became a war. For the next twelve years, the fortunes of the warring factions swung back and forth. The conservationists at first seemed to have the upper hand. When they protested the Hetch Hetchy scheme and objected to any application of the Right of Way Act in the valley, the secretary of the interior, Ethan Hitchcock, was sympathetic. "Secretary Hitchcock," wrote Colby, "relieved our apprehension and held that this act did not authorize the destruction of one of the greatest wonders of the Park." When, eight months after passage of the act, Mayor James Phelan of San Francisco, emboldened by its provisions, filed an application with the Register of the Stockton Land Office *in his own name* for reservoir rights at Lake Eleanor and Hetch Hetchy, Secretary Hitchcock denied it. When Mayor Phelan transferred to the City his interest in the two reservoir sites, the application came up for rehearing, and Hitchcock denied it again. The City then submitted a "petition for review," and for a time this sat on Hitchcock's desk.

Then, in 1902, the San Francisco electorate, tiring of Phelan and his people for other reasons, struck a blow for the conservationists and their valley, voting the mayor from office and with him the incumbent supervisors who were pushing for Tuolumne water. The ousted faction continued to work behind the scenes for Tuolumne water, but the administration of the new mayor, E.E. Schmitz, was less enthusiastic. When in February 1905 Secretary Hitchcock denied the City for the third time, rejecting Phelan's petition for review, Mayor Schmitz lost interest entirely. A year later, on February 6, 1906, the San Francisco Board of Supervisors passed Resolution 6949, asking that the City drop its plans for Hetch Hetchy and seek water elsewhere.

Then, two months later, on April 18, 1906, at 5:12 in the morning, from an epicenter two miles offshore, the San Andreas Fault struck a blow against the conservationists and in favor of the dam. The great San Francisco earthquake of 1906 destroyed the city, killed three thousand people, and brought the Tuolumne water plan back to life.

Most of the earthquake damage was not directly from the temblor, but from the fires it started, and this inspired ex-mayor Phelan to charge Secretary Hitchcock with complicity: the secretary's denial of the City's applications for reservoir sites, Phelan claimed, had dried up the fire hoses and contributed to the disaster. There was no truth to this. The fires were fed by broken gas mains all across the city, and the scarcity of water for fighting them was owed to broken distribution pipes; the conflagration had nothing to do with water sources outside the city. W.B. Acker, chief of the Patents and Miscellaneous Division of the Department of the Interior, pointed this out in a memo to Hitchcock, adding that even if the City's applications had been granted, the Tuolumne water system could not possibly have been finished until at least 1908, two years after the quake. Phelan's argument made no sense, but it had emotional appeal, and the dam proponents ran with it.

Phelan's faction continued to push for the Tuolumne water option in Washington. Marsden Manson, the San Francisco city engineer, worked with Gifford Pinchot, the first chief of the U.S Forest Service, and Benjamin Ide Wheeler, president of the University of California, among others, in the effort to bring President Roosevelt into San Francisco's camp. The enlistment of Pinchot would prove the turning point in the contest.

We complain today about the incivility of our political discourse, and we do seem to have bottomed out in one of the nadirs, but postfrontier public rhetoric was savage, too, and the exchanges in the Hetch Hetchy battle were blistering.

"I am sure he would sacrifice his own family for the preservation of beauty," Mayor Phelan said of John Muir. "He considers human life very cheap, and he considers the works of God superior."

Muir cheerfully returned fire. In an essay on Hetch Hetchy, he acknowledged that Yosemite, Yellowstone, and Sequoia National Parks were preservation triumphs, but added, "Nevertheless, like anything else worthwhile, from the very beginning, however well guarded, they have always been subject to attack by despoiling gainseekers and mischief-makers of every degree from Satan to Senators."

"He is a man entirely without social sense," Congressman William Kent said of Muir. "With him, it is me and God and the rock where God put it, and that is the end of the story."

The San Francisco Call accused the Sierra Club of something like treason. "ENEMIES HARMING THE CITY," ran a *Call* headline on December 11, 1909:

> The persistent and vicious opposition that has been directed at San Francisco in connection with the city's attempt to gain a Sierra water supply has taken fresh root within the city itself and a far reaching campaign against the acquisition of Sierra water rights is now being directed by an organization maintaining its headquarters in San Francisco. The Sierra Club, always aligned with the interests opposed to the city on the water question, is at present lending itself and its influence to an attack designed to do the city incalculable harm.

In the Hetch Hetchy battle, the language of the modern debate between preservationists and utilizers began to crystallize. The *San Francisco Chronicle* described opponents of the dam as "hoggish and mushy esthetes." The city engineer, Marsden Manson, disparaged the opposition as "short-haired women and long-haired men." Editorial cartoonists in the San Francisco papers specialized in portraying the dam's opponents as unmanly and impractical. One *San Francisco Call* artist, the young Paul Terry—he would later would move to New York, start the animation company Terrytoons, and father Mighty Mouse—drew a cartoon, "Sweeping Back the Flood," that portrayed the bearded Muir in a Mother Hubbard and apron, with daisies in his bonnet and a broom in hand, trying to sweep back a torrent labeled "Hetch Hetchy Project." This characterization of tree huggers as elitist and effete is a tradition still with us.

Hetch Hetchy was template, or at least harbinger, for one trait that has dogged the environmental movement ever since, a characteristic the movement could have done better without: a tendency toward internal division and feuding.

The most famous split came between two sometimes allies, John Muir of the Sierra Club and Gifford Pinchot of the Forest Service. The two had had an earlier falling out in 1897, when Pinchot came out in favor of sheep grazing in forest reserves and Muir responded, "I don't want anything more to do with you." Then, in 1905, Pinchot backed San Francisco in the Hetch Hetchy controversy and described a reservoir in the valley as "the highest possible use which could be made of it," at which the break with Muir became irreparable. This was not particularly lamentable, as it would have happened soon enough anyway. Pinchot was essentially a logger, a utilitarian who had brought German forestry methods to America and believed that "forestry is tree farming." Conservation for him was the husbanding of material resources. Muir was an antiutilitarian who conceded the need for lumber but believed that the great value of trees was spiritual. The two men were not subspecies of the same animal; they belonged to entirely different genera. A revision in the taxonomy of "conservationist" was overdue, and Hetch Hetchy sped that up: Pinchot's way of thinking was designated "conservationist" and Muir's "preservationist."

Hetch Hetchy also sparked feuding *within* the Sierra Club, and here it is difficult to find an upside.

The San Francisco attorney Warren Olney, a charter member of the Club and effectively its midwife—the organization took its first breaths in his office in 1892—was a strong supporter of San Francisco's position on Hetch Hetchy, and a number of other Bay Area members felt the same way. City Engineer Marsden Manson, the point man for the City in its fight for the valley, was himself a Sierra Club member. The pro-dam minority within the Club complained with growing rancor about the majority position of the Club's board. The dissension dispirited Muir, who determined to resign as president and member until Colby persuaded him to reconsider. The Olney faction leaked news of the internal controversy to *The San Francisco Call,* apparently, for in December 1909 the paper reported: "Leaders in the Sierra Club have taken emphatic exception to any attempt to commit the organization to a policy antagonistic to the City. A movement was inaugurated yesterday by which the forces favoring the city may unite and call for a poll of the Club. It is strongly hinted that a majority of the members stand ready to repudiate any action or movement that may be regarded as hostile to the interests of the city."

This proved to be wishfulness on the part of *The Call*. When the poll was finally conducted, the Sierra Club membership trounced the pro-dam faction, voting 589 to 161 in favor of continuing the fight for Hetch Hetchy. Fifty members resigned in protest, but most of the dissenters stayed. "Because," as Colby would say, "they thought they could do us more harm by saying they were members and were in favor of the Hetch Hetchy dam."

And so it has gone ever since.

The Sierra Club has been troubled for most of its history by its fifth columns, its infiltration by corporate sensibilities and influence, and by conflicts of interest among its officers. The organization does much fine work, yet it regularly steps into ethical potholes.

This has been a problem for the environmental movement as a whole. With notable exceptions—mostly in small grassroots outfits starting out—the trend in the last three decades of environmentalism, certainly in the big national organizations, has been: more MBAs as chief executives, more corporate influence on boards, more business language ("branding" and "marketing") in boardrooms and staff meetings, more environmentally suspect stock in portfolios, more of organizational budgets dedicated to fundraising and less to conservation staff, more constraint in telling truth to power and less boldness in attack.

The fight for Hetch Hetchy launched the modern environmental movement and revealed, for the first time, its surprising power and promise. At the same time it hinted at what threatens to be the movement's undoing. In the sudden bright expansion at the start there was already contraction, an impulse for regression to the mean, a retreat toward business as usual.

Hetch Hetchy had begun to slip away from the conservationists. In 1905, City Engineer Marsden Manson renewed San Francisco's application for water rights in the valley and in Lake Eleanor above. President Roosevelt determined that the secretary of the interior had the power to grant this right. That secretary had been Ethan Hitchcock, who had repeatedly denied the City's applications, but in 1907 Hitchcock resigned and was replaced by James Garfield, who saw things differently. On May 11, 1908, Garfield decided in favor of the City. "It must be remembered," he wrote William Colby the day after his decision, "that the duty imposed upon the Secretary of the Interior in acting on grants of this kind prevents him from considering merely the preservation of the park in its natural state, but he must, as well, consider what use will give the greatest benefit to the greatest number."

It was an interpretation that left the new national park system vulnerable to the whims of a cabinet officer—to a man like the former car dealer Douglas McKay, secretary of the interior under Eisenhower, or James Watt, who filled this post under Reagan—but Teddy Roosevelt, the great park maker, who was now under the sway of the formidable Gifford Pinchot, did not object.

"San Francisco against the nation for the Yosemite," ran a headline in the East Coast magazine *The World's Work*. It was true. The debate over Hetch Hetchy was indeed a national debate, and most of the republic opposed the parochial water interests of the City; the nation recognized that its interest lay in the integrity of the national parks. Nowhere was this sentiment more clearly expressed, or the arguments against the dam more succinct, than in a series of editorials in *The New York Times* in 1913, as months of Hetch Hetchy debate in Congress came to a vote.

"A national park threatened," *The Times* editorialized on July 12, 1913:

> Why the City of San Francisco, with plenty of collateral sources of water supply, should present an emergency measure to the special session of Congress whereby it invade the Yosemite National Park is one of those Dundrearian things that no fellow can find out. The Hetch Hetchy Valley is described by John Muir as a "wonderfully exact counterpart of the great Yosemite." Why should its inspiring cliffs and waterfalls, its groves and flowery, parklike floor, be spoiled by the grabbers of water and power? The public officials of San Francisco are not even the best sort of politicians; as appraisers and appreciators of natural beauties their taste may be called in question.

For an editorialist from New York, a city still run by Tammany Hall, to fault San Francisco's public officials as "not even the best sort of politicians" took gall, but it was true enough: San Francisco, like New York, was governed by crooked politicians.

"Hetch Hetchy" ran the headline on September 4, 1913:

> The only time to set aside national parks is before the bustling needs of civilization have crept upon them. Legal walls must be built about them for defense, for every park will be attacked. Men and municipalities who wish something for nothing will encroach upon them, if permitted. The Hetch Hetchy Valley in the Yosemite National Park is an illustration of this universal struggle.
>
> The House of Representatives yesterday passed a bill of the politicians of San Francisco who are nurturing a power project under the guise of providing a water supply for San Francisco. The attempt has been made to suppress a report that the Mokelumne River

would furnish a better and cheaper source than the Hetch Hetchy. The army engineers who passed favorably on the data presented to them by the officials of San Francisco— they made no investigation themselves—declared that the present water supply of the Far Western city can be more than doubled by adding present nearby sources and more economically than by going 142 miles to the Sierras.

In the tenor of *The Times* editorials, ever more anxious about the fate of the valley, a reader can sense the downward arc of the hopes of the conservationists. "THE HETCH HETCHY STEAM ROLLER," ran the headline on October 2, 1913:

> The Senate of the United States, designed by "the Fathers" to afford a wise check upon presumably impulsive action by the lower house and called "the most august deliberative body in the world," now has a chance to put a spoke in the wheel of the steam roller by which San Francisco's official lobby has heretofore crushed opposition to the Hetch Hetchy bill.
>
> The local strength behind the city's rushline is not difficult to understand when one realizes that the bill involves contracts amounting to $120,000,000, with endless opportunities of "honest graft." For months the project has been presented to Congress with persistence and specious misrepresentation. Urged first as a measure of humanity, it has been shown to be a sordid scheme to obtain electric power.
>
> The act creating the Yosemite National Park sets forth the importance and duty of reserving these wonders "in their original state," and the world has a moral right to demand that this purpose shall be adhered to. The "beautiful lake" theory deceives nobody. An artificial lake and dam are not a substitute for the unique beauty of the valley.

"ONE NATIONAL PARK GONE," *The Times* editorialized on December 9, 1913, the day after the fight ended:

> Any city that would surrender a city park for commercial purposes would be set down as going backward. Any State Legislature that would surrender a State park would set a dangerous and deplorable example. When the Congress of the United States approves the municipal sandbagging of a national park in order to save some clamorous city a few dollars, against the protests of the press and the people, it is time for real conservationists to ask, What next?

The Senate passed the Hetch Hetchy bill by a vote of 42 to 25. The bill converts a beautiful national park into a water tank for the City of San Francisco. The San Francisco advocates of the spoliation handsomely maintained at Washington, month after month, quite openly, a very competent and plausible lobbyist, and save for a few hearings and protests he occupied the Washington field most comfortably alone and unopposed. For this first invasion of the cherished national parks the people of the country at large are themselves to blame. The battle was lost by supine indifference, weakness, and lack of funds. All conservation causes in this country are wretchedly supported financially, and this one seems not to have been supported at all.

Ever since the business of nation-making began, it has been the unwritten law of conquest that people who are too lazy, too indolent, or too parsimonious to defend their heritages will lose them to the hosts that know how to fight and to finance campaigns.

With the passage of the Raker Act in 1913, work on Hetch Hetchy Dam and its delivery system began. The project brought a new gold rush to the western slope of Yosemite and the foothills below, the country where the original gold rush had petered out a half century before. The ghost town of Groveland returned to life as a staging area: homes, hospital, schools, shops, saloons. Quarries were dug in Tuolumne Canyon for granite blocks and riprap, and a sawmill built. Hetch Hetchy Valley was logged, and the timber dragged by mule team to the mill and sawn into planks and beams for buildings and barracks and framing and supports. The mill churned out ties for the sixty-eight miles of the Hetch Hetchy Railroad, which climbed eastward from the foothills into the mountains, transporting materials and thousands of workers to the canyon of the Tuolumne.

To provide electricity, a dam twelve hundred feet long and seventy feet tall was built on Lake Eleanor, to the northeast—sited, as the main dam would be, within the borders of the national park. The deepened lake's water was directed through tunnels, flumes, and canals to a hydro plant at Early Intake, downstream from Hetch Hetchy. From Early Intake, transmission lines radiated outward, electrifying the reviving town of Groveland and the various quarrying and construction sites. Access roads were built into the park, among them a steep, winding gravel road to Lake Eleanor, and a gentler, wider, graded roadbed into Hetch Hetchy Valley. At the damsite, Hetch Hetchy Camp went up: first a dining hall seating five hundred workers, then a bunkhouse, a clinic, and warehouses for wood, oil, and meat.

Lake Eleanor hydroelectricity powered drills that bored a diversion tunnel a thousand feet long and twenty feet wide down through the hard granite of the valley wall. A cofferdam was built across the river, diverting the Tuolumne past the damsite into the tunnel and leaving the river bottom dry for construction. Here at the narrows the channel was chocked 120 feet deep with detritus—river boulders, stones, sand—which were removed to expose bedrock. The smooth granite of the opposing walls was roughened for adhesion with the concrete of the dam. The concrete was "cyclopean," a modern imitation of the cyclopean masonry of ancient Greece, Crete, and Italy, which joined enormous blocks of stone without mortar, a technique that originated, the story goes, with the Cyclopes, the Thracian tribe of one-eyed giants. In the modern version, the "plums" or "pudding stones" of the blocks are set in a matrix of concrete—in the case of Hetch Hetchy, blocks of Sierra granite as big as six cubic yards.

Hetch Hetchy Dam rose tier by tier of cyclopean blocks, not entirely unlike the Citadel of Tiryns or Machu Picchu, and began to take shape across the narrows at the foot of the valley. The shape it took was the graceful curve of the arch-gravity dam. An arch-gravity dam combines the virtues of the gravity dam and the arch dam. A gravity dam, Grand Coulee for example, by its sheer mass and the gravity acting upon that mass, resists the weight of water trying to push it over. An arch dam curves its convex side upstream, which directs the water pressure laterally against the opposing walls of the canyon, compressing and strengthening the arch. An arch-gravity dam does a lot of both. It requires less internal fill than a pure gravity dam and can be made thinner and more elegant. There is beauty in the arch, as the Moors and Romans discovered in their architecture. In a nondescript or ugly canyon, if such a place exists, the arch-gravity dam might be seen as ornament, as an improvement. O'Shaughnessy Dam's designers and the City of San Francisco certainly saw it that way.

As the dam climbed to fill the gap in the Hetch Hetchy narrows, work went on simultaneously on the sections downstream. The diversion tunnel at the damsite, a thousand feet long, was the smallest of the bores that would connect the reservoir with San Francisco. From Early Intake, the captured Tuolumne would flow through the nineteen miles of Mountain Tunnel to the top of Priest Grade, then the sixteen miles of the Foothill Tunnel from Moccasin Reservoir to Oakdale Portal, then the forty-seven miles of pipeline from Oakdale across the San Joaquin Valley, then the twenty-nine miles of the Coast Range Tunnel to the shore of San Francisco Bay, and finally the twenty-five miles of underwater pipeline to Crystal Springs Reservoir.

The most heroic and difficult work of the project was not on the dam itself, but underground in these endless tunnels connecting to the coast. The Mountain Tunnel ran, on average, a thousand feet underground, through fissured rock that dripped and drizzled everywhere, forcing the men to work in oilskins like sailors in a storm, with huge pumps draining thousands of gallons a minute from the tunnel to keep them from drowning. The Coast Range Tunnel was a particular worry to consulting engineers, who predicted troubles with groundwater, groundswell, and gases. They advised that on this stretch the Tuolumne be pumped in pipes over the Coast Range, instead of running through gravity-flow tunnels under it. Michael O'Shaughnessy, the chief of the project and head engineer, insisted that a pipeline would cost more to build and would carry less volume than his tunnels. He pointed out that gravity flow, which is free, would save all the electric costs of pumping water over the Coast Range for the lifetime of the system.

O'Shaughnessy's tunnels were a John Muir nightmare. In *My First Summer in the Sierra,* Muir wrote of the destruction brought to the Sierra foothills by miners, "especially in the lower gold region—roads blasted in the solid rock, wild streams dammed and tamed and turned out of their channels and led along the sides of cañons and valleys to work in the mines like slaves." O'Shaughnessy saw it differently, of course. In the matter of river slavery, the engineer was no abolitionist. If cost-saving was part of his motivation for the tunnels, then another was the simplicity and purity of gravity flow. Gravity is what powered the Assyrian, Greek, Roman, and Incan aqueducts. There is a working aqueduct in northern Iraq that has been continuously delivering its water gravitationally since seven hundred years before Christ. Muir's was not the only sort of poetry possible in these mountains. O'Shaughnessy had a classical aesthetic of his own.

The auguries of the critics of the Coast Range Tunnel proved to be on target. At a depth of twenty-five hundred feet under Crane Ridge at Livermore, groundswell constricted, in a single day, a bore with a diameter of eighteen feet to a passage too tight for a human to wiggle through. The tunnel was rebored larger, reinforced with thick rings of gunite, and then lined with concrete. In July 1931, one crew deep under the Coast Range encountered methane, another worry of the critics, and the explosion killed twelve men. O'Shaughnessy's tunnels were not cost-free, then, but workers died aboveground, too, in building the Hetch Hetchy system.

An irony of the Hetch Hetchy project is how few of the protagonists saw it through to conclusion. Muir died twenty years before completion, never having had to witness any changes to his valley. He was "heartbroken"—the word almost every Muir biographer chooses—by passage of the Raker Act the year before his death, yet he also expressed relief that the long fight was finally over. James Phelan died in 1932, two years too soon to see his dream for Tuolumne water realized. Michael O'Shaughnessy died on October 12, 1934, just sixteen days before his system was finished and dedicated. William Mulholland, who helped his colleague J.B. Lippincott design the Hetch Hetchy pipeline across the San Joaquin Valley, lived on nine months after the dedication, but he was no longer a player. Mulholland, even as he worked on the Hetch Hetchy pipeline, had been directing his Los Angeles Department of Water and Power in its construction of the St. Francis Dam—an arch-gravity structure, like Hetch Hetchy—in the mountains northwest of Los Angeles. In 1928, just hours after Mulholland gave the St. Francis Dam a personal safety inspection and awarded it a thumbs-up, the dam failed, killing between four hundred and six hundred people downstream and ending Mulholland's career.

O'Shaughnessy Dam and the gravity-flow system of tunnels and pipelines that delivered Tuolumne water to San Francisco were an engineering marvel. Many understood at the time, and historians are now nearly unanimous in retrospect, that it need not have been so marvelous. There were better, smarter, simpler, lower places for San Francisco to have stored its Sierra Nevada water. Ingenuity is often like this; in rising to an engineering challenge, it sometimes transcends common sense. For the past eighty years, the dam at Hetch Hetchy has stood symbolic, a temple of elegant solution or a monument of inappropriate technology, whichever you want.

The Call=Chronicle=Examiner

SAN FRANCISCO, THURSDAY, APRIL 19, 1906.

EARTHQUAKE AND FIRE: SAN FRANCISCO IN RUINS

Background: *The San Francisco Call, Chronicle,* and *Examiner* newspapers combined the day after the earthquake to describe the destruction. *Courtesy of the Doe Library, University of California, Berkeley*

The fire rages full force at Third and Mission Streets. *Courtesy of the California History Room, California State Library, Sacramento, California*

Broken water pipes hampered firefighting efforts in San Francisco following the earthquake. *Courtesy of The Bancroft Library, University of California, Berkeley, BANC PIC 1958.021 Vol. 1: 154—fALB*

"OUR LATE EARTHQUAKE AND FIRE HAS OPENED THE EYES TO THE FACT THAT WE HAVE THE POOREST WATER-SYSTEM OF ANY CITY IN THE UNITED STATES," COMMENTED ENGINEER AUGUSTUS WARD TO ROBERT UNDERWOOD JOHNSON IN 1908. THE LONG-TABLED HETCH HETCHY PLAN FOUND RENEWED SUPPORT.

Flames consume a building on California Street while a fire engine, lacking water, is useless. *Courtesy of the Huntington Library, San Marino, California*

The front page of *The San Francisco Call* newspaper three days after the earthquake. *Courtesy of the Doe Library, University of California, Berkeley*

"WATER IN SAN FRANCISCO COSTS MORE THAN BREAD, MORE THAN LIGHT
SPEAKING OF SAN FRANCISCO'S PRIVATE WATER SUPPLIER
SAN FRANCISCO MAYOR JAMES D. PHELAN FILED TO OBTAIN WATER
WITH ENGINEERING INGENUITY, CLEAN SIERRA WATER COULD BE

This editorial cartoon from June 1908 captured San Francisco's frustration with the Spring Valley Water Company. *Courtesy of The Bancroft Library, University of California, Berkeley, BANC MSS 71/103 C Vol. 5*

San Francisco Mayor James D. Phelan. *Courtesy of the San Francisco History Center, San Francisco Public Library*

OMPLAINED POLITICAL ECONOMIST HENRY GEORGE,
PRING VALLEY WATER COMPANY. **AS EARLY AS 1901,**
IGHTS IN HETCH HETCHY, THE GRANITE-WALLED VALLEY WHERE,
ARNESSED FOR THE GROWING METROPOLIS.

The outlet of the Tuolumne River at Hetch Hetchy Valley presented a perfect damsite. *Courtesy of The Bancroft Library, University of California, Berkeley, BANC PIC 1932.001.102.043—ALB*

"VOTE FOR HETCH HETCHY AND A FREE CITY"
— *The San Francisco Call*, November 12, 1908

The front page of *The San Francisco Call* the day before the city's November 1908 special election on the Hetch Hetchy water project. *Courtesy of The Bancroft Library, University of California, Berkeley, BANC MSS 71/103C Vol. 5*

LADY SAN FRANCISCO BEGAN TO BREAK OGRE-ISH SPRING VALLEY WATER COMPANY'S HOLD IN NOVEMBER 1908 WHEN VOTERS BESTOWED THE GRAIL OF HETCH HETCHY UPON HER— YET IT WOULD STILL BE FIVE YEARS BEFORE THE FEDERAL GOVERNMENT FINALLY APPROVED THE PROJECT.

A large number of San Francisco voters turned out to approve a $600,000 Hetch Hetchy bond issue. *Courtesy of The Bancroft Library, University of California, Berkeley, BANC MSS 71/103C Vol. 5*

greatly enhanced.

Ever since the Park was established it has called for defense, much it may be invaded or its boundaries shorn while a single mountain or tree or waterfall is left the poor stub of a park would still need protection. The first forest reserve was in Eden and though its boundaries were drawn by the Lord, and angels set to gaurd it, even that most moderate reservation was attacked.

I pray therefore that the people of California be granted time to be heard before this reservoir question is decided: for I believe that as soon as light is cast upon it, nine tenths or more of even the citizens of San Francisco would be opposed to it. And what the public opinion of the world would be may be guessed by the case of the Niagara Falls.

Faithfully and devotedly yours

John Muir

The President—

O for a tranquil camp hour with you like those beneath the Sequoias in memorable 1903.

In this 1907 letter, Muir urged President Theodore Roosevelt to preserve Hetch Hetchy Valley, and added a handwritten note that recalled the Yosemite camping trip they took together in 1903. *[muir16_0986-md-1] John Muir Papers, Holt-Atherton Special Collections, University of the Pacific Library ©1984 Muir-Hanna Trust*

"HETCH HETCHY VALLEY MENACED—SAN FRANCISCO'S SELFISHNESS WOULD ROB STATE OF ONE OF ITS MOST BEAUTIFUL SPOTS, RESERVED IN NATIONAL PARK, THE PEER OF YOSEMITE"

—ARTICLE BY JOHN MUIR, UNKNOWN PUBLICATION, C. 1908

> "AS A FIGHTER FOR NATURE, MR. MUIR IS IN ALL HIS GLORY, AND HE IS FIGHTING WITH EVERY POWER HE CAN COMMAND WHAT HE TERMS THE DESECRATION OF HETCH HETCHY VALLEY BY SAN FRANCISCO."
> —*Los Angeles Evening Express*, early 1900s

This early 1900s article in the *Los Angeles Evening Express* newspaper publicized Muir's fight to save Hetch Hetchy. *Courtesy of The Bancroft Library, University of California, Berkeley, BANC MSS 71/103, Vol. 6*

Fearing that a bill to approve the dam would be rushed through Congress, Muir sent this urgent 1913 telegram to Robert Underwood Johnson, an East Coast editor and friend who lobbied for protection of the valley. *Courtesy of The Bancroft Library, University of California, Berkeley, BANC MSS C-B 385, box 4*

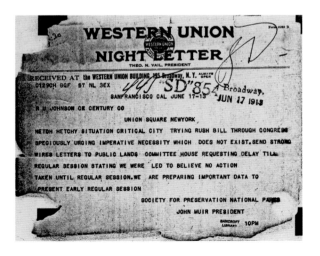

John Muir was mocked in this 1909 *San Francisco Call* front page editorial cartoon.
Courtesy of the Doe Library, University of California, Berkeley

The *Los Angeles Times* described Muir as "Nature's Champion." *Courtesy of The Bancroft Library, University of California, Berkeley, BANC MSS 71/103 C, Vol. 7*

FLOWERY, GIRLISH, AND NAÏVE— OR STALWART DEFENDER OF BEAUTY?

NATURE'S CHAMPION.

IN LABOR OF LOVE FOR A BEAUTY PET.

John Muir, Noted Lover of Nature, who says San Francisco wants to trade a hog ranch for the Hetch-Hetchy Valley, one of God's beautiful flower gardens.

SAN FRANCISCO AGAINST THE NATION FOR THE YOSEMITE

AN ATTEMPT OF THE CITY TO ACQUIRE A PART OF A NATIONAL PARK FOR A WATER SUPPLY

The World's Work magazine, April 1909

> "THE ACT OF CREATING THE YOSEMITE NATIONAL PARK SETS FORTH THE IMPORTANCE AND DUTY OF RESERVING THESE WONDERS 'IN THEIR ORIGINAL STATE' …AN ARTIFICIAL LAKE AND DAM ARE NOT A SUBSTITUTE FOR THE UNIQUE BEAUTY OF THE VALLEY."
> — THE NEW YORK TIMES, OCTOBER 2, 1913

Harper's Weekly magazine featured this article in September 1909. *Courtesy of The Bancroft Library, University of California, Berkeley, BANC MSS C-B 416*

"THE CITY IN COURSE OF ITS DEVELOPMENT OF WORKS FOR WATER SUPPLY WILL MAKE THE HETCH HETCHY VALLEY MORE BEAUTIFUL, AND A FAR MORE USEFUL INSTRUMENT OF PLEASURE THAN IT IS TODAY."

—John Freeman, engineer

John Freeman, a prominent engineer, wrote an extensive report for San Francisco outlining the benefits of building a dam at Hetch Hetchy. The 1912 report included stylized pictures of the future reservoir, including the image above, and proposed a scenic road around the entire lake, which was never built. *Courtesy of The Bancroft Library, University of California, Berkeley, BANC MSS 71/103 C, Vol. 7*

On December 2, 1913, as the U.S. Senate began to discuss the Hetch Hetchy bill, *The San Francisco Examiner* published a sixteen-page special edition, distributed in Washington, D.C., which enumerated the advantages of the proposed reservoir. The edition included this full-page editorial, which suggested that Hetch Hetchy could be more beautiful as a lake. *Courtesy of* Hetch Hetchy; the story of San Francisco's struggle to provide a water supply for her future needs *by Ray W. Taylor (Orozco 1926)*

San Francisco Chronicle,
September 1908

The San Francisco Examiner ran a Hetch Hetchy editorial cartoon on December 5, 1913, the day before the Senate approved the flooding of the valley. Nature-loving John Muir look-alikes and those opposed to the dam, including Senator John Works, were portrayed as being on the run. *Courtesy of the Doe Library, University of California, Berkeley*

IN DECEMBER 1913, THE U.S. SENATE VOTED TO APPROVE THE DAM AND PRESIDENT WOODROW WILSON SIGNED THE RAKER ACT. "THE BATTLE WAS LOST BY SUPINE INDIFFERENCE, WEAKNESS, AND LACK OF FUNDS," LAMENTED AN EDITORIAL IN *THE NEW YORK TIMES*.

SAN FRANCISCO WINS HETCH HETCHY; VOTE 43 TO 25

Battle for Water Rights Comes to Triumphant Finish in United States Senate

The front page of *The San Francisco Examiner* the day after the Senate voted to approve the reservoir. *Courtesy of the Doe Library, University of California, Berkeley*

Upon signing the Raker Act, President Woodrow Wilson presented the pen he used along with an official memo to San Francisco Mayor James Rolph, Jr. *Courtesy of the Yosemite NPS Library*

San Francisco officials joyride while defeated preservationists react to their injuries in this Chopin cartoon on Hetch Hetchy politics. *Courtesy of The Bancroft Library, University of California, Berkeley, BANC PIC 1992.058-PIC—AX*

John Muir in his older years. *[MSS048.f23-1270] John Muir Papers, Holt-Atherton Special Collections, University of the Pacific Library* ©1984 Muir-Hanna Trust

"THE DESTRUCTION OF THE CHARMING GROVES AND GARDENS, THE FINEST IN ALL CALIFORNIA, GOES TO MY HEART. BUT IN SPITE OF SATAN & CO. SOME SORT OF COMPENSATION MUST SURELY COME OUT OF THIS DARK DAMN-DAM-DAMNATION."

—JOHN MUIR

This part of the Valley will soon be a reservoir for San Francisco water supply.

A group of Campfire Girls visited Hetch Hetchy in 1919 and included this image in a photo album. *[Mss84]*, *Holt-Atherton Special Collections, University of the Pacific Library*

City Engineer Michael O'Shaughnessy turned this pre-dam photograph of Hetch Hetchy Valley—taken from the road to Lake Eleanor—into a Christmas card. *Courtesy of The Bancroft Library, University of California, Berkeley, BANC PIC 1992.058—PIC, copy 2, Ctn 3:43*

Although the valley still provides a scenic background for this early 1900s photograph, the railroad tracks signal that change is ahead. *Courtesy of The Bancroft Library, University of California, Berkeley, BANC PIC 1992.058-PIC, Ctn 3:59*

A tour group takes in the view of the famous valley that will soon be a reservoir, c. 1920. *Courtesy of the Yosemite NPS Library*

An August 1919 view of the valley floor. *Courtesy of The Bancroft Library, University of California, Berkeley, BANC PIC 1992.058—PIC Ctn 3:43*

Hetch Hetchy Valley, after trees were removed, looking upstream from the north side of the valley. *Courtesy of the San Francisco History Center, San Francisco Public Library*

Looking upstream from the south side, April 22, 1922. *Courtesy of The Bancroft Library, University of California, Berkeley, BANC PIC 1992.058—PIC, Ctn 3:70*

Hetch Hetchy Aqueduct Tunnel Heading, 1921. *Courtesy of The Bancroft Library, University of California, Berkeley, BANC PIC 1992.058-PIC, Box B*

September 1922. *Courtesy of The Bancroft Library, University of California, Berkeley, BANC PIC 1992.058—PIC, Ctn 3:73*

Michael O'Shaughnessy, the head of dam construction (right), with workers. *Courtesy of The Bancroft Library, University of California, Berkeley, BANC PIC 1992.058—PIC, Ctn 5:40*

Installing machinery at the Moccasin Power Plant, November 1924. *Courtesy of The Bancroft Library, University of California, Berkeley, BANC PIC 1992.058—PIC, Ctn 4:57*

Workers laying submarine pipeline across the Newark Slough in February 1925. *Courtesy of the San Francisco History Center, San Francisco Public Library*

Raising the height of the dam and increasing reservoir storage capacity, July 1937. *Courtesy of the Yosemite NPS Library*

Steel anchor bars between the old and the new concrete, March 1937. *Courtesy of the San Francisco History Center, San Francisco Public Library*

The Groveland roundhouse crew for the Hetch Hetchy Railroad, May 1922. *Courtesy of the San Francisco History Center, San Francisco Public Library*

Beautiful color prints were produced of the finished dam. *Courtesy of The Bancroft Library, University of California, Berkeley, BANC PIC 1992.058-PIC, Ctn 3:85*

Hetch Hetchy Valley, viewed from the trail to Lake Eleanor, 1906. *Courtesy of the U.S. Geological Survey, Department of the Interior/USGS*

The same view of the valley after it was flooded, c. 1923. *Courtesy of The Bancroft Library, University of California, Berkeley, BANC PIC 1992.058-PIC, Ctn 3:83*

San Francisco officials, including Mayor James Rolph (right) and City Engineer Michael O'Shaughnessy (center), gathered in 1923 to celebrate the completion of O'Shaughnessy Dam.
Courtesy of The Bancroft Library, University of California, Berkeley, BANC PIC 1992.058—ALB v. 4:30a

City Engineer Michael O'Shaughnessy featured the new dam on his annual Christmas card in 1923.
Courtesy of The Bancroft Library, University of California, Berkeley, BANC PIC 1992.058-PIC, Ctn 7

An elaborate multipage program was produced for the 1934 celebration of the first delivery of Hetch Hetchy water. *Courtesy of the San Francisco History Center, San Francisco Public Library*

Colored postcards from the era emphasized the beauty of the dam. *Courtesy of The Bancroft Library, University of California, Berkeley, BANC PIC 1932.001—AL, Vol. 94*

Photographer Philip Hyde captured low-water conditions at the reservoir in 1955.
Field of Stumps, Hetch Hetchy Valley, Yosemite National Park, California. ©*Philip Hyde, 1955*

The *Stockton Record* newspaper ran this editorial cartoon in July 1928 in response to news that San Francisco did not plan to build the promised road around the reservoir and was concerned about pollution from greater public access. *Courtesy of the Yosemite NPS Library*

THOUGH THE HETCH HETCHY SYSTEM HAS DELIVERED HIGH-QUALITY DRINKING WATER TO MILLIONS OF PEOPLE IN THE BAY AREA FOR DECADES, THE RESERVOIR IS NEITHER AS ACCESSIBLE NOR AS BEAUTIFUL, ESPECIALLY DURING LOW-WATER CONDITIONS, AS SAN FRANCISCO LED THE AMERICAN PEOPLE TO BELIEVE IT WOULD BE.

AGE OF DAMS 3

"**The twentieth century** has been the Hydraulic Century, the Age of Dams," water historian Marc Reisner wrote in 1999, as this strange period gave its last gasp, or gurgle, and came to a close. "At least 95 percent of mentionable dams—usually defined as those more than fifteen meters high—were built in the past hundred years."

By the end of the twentieth century, 68,000 large dams and 75,000 small ones had brought the rivers of America to a standstill. In all the lower forty-eight, only two rivers more than 120 miles long, the Yellowstone and the Salmon, flowed freely for their entire length. Our greatest river, the Mississippi, obstructed by 3,717 dams, no longer resembled the Big Muddy piloted by Mark Twain a century before. ("The Mississippi River will always have its own way; no engineering skill can persuade it to do otherwise," Twain prophesied. For once he failed to clear his namesake two fathoms. His prognostication ran aground on the bar.) East of the Mississippi, no stream of any size ran its natural course. In the South, in particular, rivers were not just dammed but catheterized, rendered sterile by the grim surgery that the Army Corps of Engineers called "channelization."

Nowhere was overdevelopment of water more dramatic than in the West. Out beyond the hundredth meridian, the great monuments to the Age of Dams rose up. The most famous of these were make-work projects, colossi for which the Public Works Administration played pharaoh. The government of the United States, in the depths of the Depression, simultaneously built the five biggest structures in the history of mankind: Hoover, Grand Coulee, Bonneville, Shasta, and Fort Peck Dams. The engineers took the model of O'Shaughnessy, the arch-gravity dam, and scaled it up vastly to make Hoover and Shasta. For good measure they scaled up O'Shaughnessy

Hetch Hetchy Dam looking northwest, by Brian Grogan. *Courtesy of the Prints and Photographs Division, Library of Congress, HAER CAL,55-MATH.V,1--7* itself, from its original 227 feet to 312, finishing the job in 1938, the year work on Hoover began. The proliferation of great dams in the West was not senseless—it did some good—but what these impoundments accomplish best, in this driest region in the nation, is evaporation of a precious resource.

In the fifth millennium, or the fiftieth, or whenever it is that the aliens set down in their spacecraft, their archaeologists will find these tall tombstones to dead rivers everywhere. The aliens will surmise, correctly, that the semi-intelligent inhabitants were in the grip of some mania. They will deduce, as archaeologists always do when confronted by monumental ruins they cannot understand, that the meaning must have been religious. Alien poets will be drawn to the subject. In Tralfamadorean, or Titanic, or some other extraterrestrial language, an alien Percy Bysshe Shelley will write the alien "Ozymandias."

    And on the pedestal these words appear:
   "My name is Ozymandias, king of kings:
   Look on my dams, ye Mighty, and despair!"

Or perhaps not.

It may not be that way at all. Toward the end of the Hydraulic Century, at the conclusion of the Age of Dams, a new trend surfaced. The alien archaeologists may find riverscapes entirely different from those we see today.

In the mid-1990s, a movement for dam removal came alive and rapidly gained momentum. This owed in small part to renewed courage in people like my father, and to analysis and sharp commentary by critics like Marc Reisner, and to the maturation of adolescent organizations like Restore Hetch Hetchy, Glen Canyon Institute, Friends of the River, International Rivers, American Rivers, the Atlantic Salmon Federation, Trout Unlimited, and the like. But it owed more to the senescence of dams. As the great gray edifices of the dam-building craze grow old, all the predictions of environmentalists and river lovers are coming true: safety and maintenance issues are multiplying, relicensing of old dams is becoming more and more difficult, reservoirs are filling up with sediment, and all the natural services of free-flowing rivers are now widely recognized and missed.

The first major dam removed for the purpose of river restoration was Edwards Dam on the Kennebec River in 1999. With the barrier of this Maine dam demolished, striped bass, alewives, and other sea-run fish are now prospering in the seventeen miles of restored habitat along the Kennebec. Prospering too are the ospreys and bald eagles that have reappeared to hunt those fish. In 2002, Mill Pond Dam was removed from the Chippewa River for reasons of safety and to restore stream habitat for bluegill, steelhead, and other native fish. In 2006, Potagannissing Dam on the Potagannising River in Chippewa County, Michigan, was removed to improve fish passage, particularly that of the northern pike.

In September 2011, in Washington state, bulldozers began chipping away at the concrete of Elwha Dam in Olympic National Park. This 108-foot-tall dam on the Elwha River, and a 210-foot companion dam upstream at Glines Canyon, were built in 1913, the year the Raker Act authorized Hetch Hetchy Dam. These two Olympic dams, like the dam in Yosemite, compromise a national park. The Elwha is one of the few West Coast rivers to host all five species of Pacific salmon: coho, Chinook, chum, pink, and sockeye, as well as four species of anadromous trout. (*Elwha,* may derive from the Klallam word *elkwah,* "elk," but it is difficult to know for sure. Only four native speakers of Klallam remain to debate the matter.) A century ago, before the dam, the Klallam had a village on the riverbank and saw runs of four hundred thousand salmon annually, the fish thronging past the village to spawn in the gravels of more than seventy miles of river. The dam reduced spawning habitat to fewer than five miles of river below its barrier, and in a good year four thousand salmon show up. The demolition on the Elwha River is the largest dam removal project ever launched in the United States. The work will require two or three years, because twenty-four million cubic yards of sediment have accumulated behind the dam. River engineers estimate that in thirty years or so the Elwha will have returned to its natural state, and to the salmon and trout.

On hand as the bulldozer went to work on Elwha Dam were Washington's governor, Chris Gregoire; the state's senators, Patty Murray and Maria Cantwell; several leaders of the Lower Elwha Klallam Tribe; and a couple of emissaries from the United States, Ken Salazar, secretary of the interior, and Michael Conner, commissioner of the U.S. Bureau of Reclamation. The battering by bulldozer amounted to a de-christening—the champagne imploding back into the bottle, the shattered glass reforming, the magnum swinging away intact from the face of the dam. "This is just breathtaking for me," said Commissioner Conner. "This is not only an historic moment, but it's going to lead to historic moments elsewhere across the country."

It amounted to an epitaph. Commissioner Conner could hardly have delivered a more ringing pronouncement of the death of the Age of Dams. Conner's agency, the Bureau of Reclamation, is the keeper of dams in the West. A student of Western water wars, on reading his words that day, suffers a moment of disequilibrium. The student remembers—or this student did, anyway—Conner's most famous predecessor, Commissioner Floyd Dominy, godfather to Glen Canyon Dam and the bureaucrat who presided over the final binge of American dam-building in the last half of the Hydraulic Century. The student tries to imagine Commissioner Conner's endorsement of the Elwha Dam demolition, and his cheerful anticipation of more demolitions to come, as coming from the mouth of Commissioner Dominy (or "Dominate," as the boss was sometimes known at the Bureau of Reclamation). Not possible. Unimaginable. Who would have ever thought?

In some watersheds, small dams have been falling like dominoes. In 2000, on Ashland Creek in Oregon, Unnamed Dams 1, 2, and 3 came down. That same year, on Muddy Run in Pennsylvania, Amish Dams 1, 2, 3, and 4 were removed, followed in 2001 by Amish Dams 5, 6, 7, and 8. In 2002, in Wisconsin, Silver Springs Dams 1, 2, 3, 4, 5, 6, 7, 8, 9, 10, 11, 12, and 13 all came down serially on one tributary to the Onion River.

In 2000, dams were removed from the Kissimmee and Muskegon Rivers and from Cocalico and Koshkonong Creeks. In 2001, dams were removed from Mahantango and Conodoquinet Creeks, in 2002 from the Chatanika, the Ashuelot, the Olentangy, and the Huron, in 2003 from the Ompompanoosuc and Ottawa, in 2004 from the Cuyahoga, the Cahaba, the Contoocook, the Maunesha, the Oconto, and the Rappahannock. In following years, dams were removed from the Waupaca, the Neshannock, the Potomac, the Musconetcong, the Mukwonago, and the Shiawassee.

Isn't it interesting how aboriginal names, even as they wash away from the general landscape, stick so stubbornly along stream courses? And isn't it an especially fine thing when those aboriginal rivers flow free again?

No really big dams have yet come down. Certainly none as large as O'Shaughnessy, the 312-foot dam inundating the Sierra valley named by the Natives for its tall grass. For the first fifty years of the dam's history, talk flared now and again among environmentalists about removing the dam and restoring

Hetch Hetchy, but never with much hope or conviction. Then in August 1987, to the surprise of almost everyone, Secretary of the Interior Donald Hodel proposed a study on decommissioning the dam and draining the reservoir. With this startling, bolt-from-the-blue suggestion from the antienvironmental administration of Ronald ("If you've seen one redwood, you've seen them all") Reagan, the Hetch Hetchy controversy returned to life and has been roiling ever since.

The debate quickly found its old channel, just as Secretary Hodel predicted the Tuolumne River would once the dam was gone.

San Francisco politicians, like their predecessors of the early 1900s, came out ardently in favor of O'Shaughnessy. Politicians beyond the borders of San Francisco Bay, like their own predecessors, tended to take another view. U.S. Senator Dianne Feinstein, a former mayor of San Francisco, distinguished herself immediately as the most formidable of modern advocates for the dam. In October 1987, two months after Secretary Hodel's bombshell, she and the interior secretary faced off atop the dam. Hodel wore jeans and cowboy boots, Feinstein a burgundy dress. It was like a duel from an old *Gunsmoke* episode, Doc Adams confronting Miss Kitty in the middle of Main Street.

"Yosemite National Park is America's birthright, not any individual city's," said Hodel.

Feinstein countered by asking how San Francisco was going to replace the pure Hetch Hetchy water upon which millions of Bay Area residents depended. She brought up the question of mosquitoes. If the reservoir were drained, said the senator, then Hetch Hetchy would revert to a swamp swarming with bloodthirsty insects. This pestilential swamp-and-mosquito imagery had worked in the early 1900s for James D. Phelan, Feinstein's predecessor as both San Francisco mayor and U.S. senator. (The Panama Canal was being dug at the time, thousands of Canal Zone workers were contracting yellow fever and malaria, and the mosquito as vector was on everyone's mind.) DEET had since been invented, as Secretary Hodel pointed out, but in the late 1900s Senator Feinstein gave mosquitoes another try.

Twenty years after Hodel's modest proposal, the debate had begun to calm slightly; then in December 2011 another Republican, the right-wing California congressman Dan Lungren, stirred the reservoir waters again by seconding the modest proposal. Lungren, who represents several foothill counties at the border of Yosemite Park, pointed out that San Francisco gets bargain-basement prices for its national park water, neglects three possible alternative sources—water recycling, groundwater, and rainwater—and could store its Tuolumne water farther downstream in other reservoirs.

These were not new ideas. In a 2004 report called "Paradise Regained," water specialists at Environmental Defense proposed a plan by which San Francisco would continue to use Tuolumne River water for almost all of its needs, mostly using existing infrastructure. The key component would be the storage that the San Francisco Public Utility Commission already owns in Don Pedro Reservoir, a capacity nearly double what the city stores in Hetch Hetchy. An intertie would have to be built to connect Don Pedro and the SFPUC's San Joaquin pipelines, but otherwise the Environmental Defense plan would make use of Michael O'Shaughnessy's fine existing gravity-flow aqueduct. San Francisco could continue to take most of its remarkably pure water supply from high up on the Tuolumne River, making its "run-of-the-river" diversions a few miles downstream of Hetch Hetchy at the Early Intake Diversion Dam, exactly where it diverted all its water to the Bay Area until 1967.

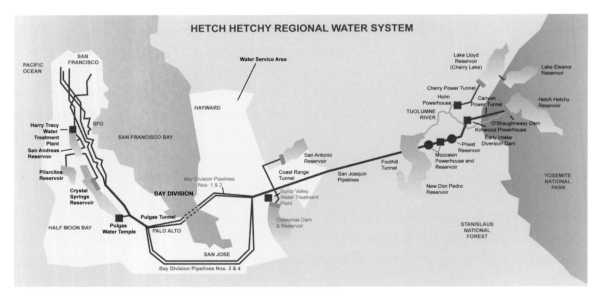

Map provided by Hetch Hetchy Regional Water System operated by the San Francisco Public Utilities Commission

When Congressman Lungren entered the fray, Senator Feinstein jumped back in herself. In *The Sacramento Bee,* in an op-ed directed at Lungren, she concluded, "To remove or in any way interfere with Hetch Hetchy's continued operation simply makes no sense." Ed Lee, the city's current mayor, agreed a month later. "As insane as this is, it is, in fact, insane," he said.

The old division over Hetch Hetchy, which pitted San Francisco against most of the rest of the nation, resurfaced in journalism, too. The *San Francisco Chronicle,* the last masthead standing of the fleet of San Francisco journals that had favored the dam at inception, renewed its support for O'Shaughnessy. Journalists elsewhere in the republic, as before, took delight in sticking it to San Francisco.

The New York Times picked up its old Hetch Hetchy themes as if it had never dropped them. "Nearly a century ago, Congress rashly approved a dam and an eight-mile-long reservoir called Hetch Hetchy in the northwest corner of Yosemite National Park," the paper recalled, in a February 16, 2012, editorial. "The project, completed in 1923 despite a national outcry, included a sweetheart deal for San Francisco: the right to buy the water for $30,000 a year." *The Times* mentioned Congressman Lungren's request that the secretary of the interior investigate San Francisco for not fulfilling its end of the Raker Act, which required the city to use all its local water sources before turning to Hetch Hetchy. "The issue is not only how San Francisco uses its water," concluded *The Times,* "but whether there's any rationale, a century later, for the dam and the deal. Ninety-nine years ago, this newspaper argued against the construction of the dam. We lost, but the much bigger loss was nature's."

Dan Morain of *The Sacramento Bee* weighed in on the subject the day before *The Times* did. For Morain, as for other California journalists outside San Francisco, the renewed Hetch Hetchy controversy was license for the kind of rambunctious, tongue-in-cheek journalism that Mark Twain had practiced in the state and that still marks Western reporting.

Morain managed to poke some fun at his brethren in the Big Apple ("As the Reagan administration was drawing to a close, Interior Secretary Donald Hodel leaked his desire to drain Hetch Hetchy to, who else, *The New York Times,* which gushed about the grand possibilities of restoring wildland 3,000 miles from its Manhattan headquarters.")

Morain poked fun at Congressman Lungren ("Lungren has never much been associated with the environment, but he faces a well-funded Democratic challenger, Ami Bera, in a swing seat this year. 'Trying to remedy one of the worst environmental travesties in California knows no partisan bounds,' Lungren's campaign aide, Rob Stutzman, told me with a straight face, more or less.")

Morain poked fun at San Francisco, which faced a November initiative put on the ballot by the organization Restore Hetch Hetchy ("No city in America prides itself on its environmental credentials more than San Francisco. If the initiative goes forward, Hetch Hetchy Reservoir defenders could have a hard time explaining to some voters why they shouldn't support a process that would lead to the dismantling of O'Shaughnessy Dam. Lungren would be happy to pitch in if environmentalists need help swaying any Republicans still in San Francisco.")

The editorial cartoonist at the *San Francisco Chronicle*, Tom Meyer, inspired by the recent sighting of a wolf in the northern part of the state—the first California wolf since 1924—portrayed Dan Lungren at Hetch Hetchy as a wolf in sheep's clothing. The wolf grins toothily, holds up a "Restore Hetch Hetchy" placard, and flashes the peace sign at a flock of sheep. One sheep says, "He looks and sounds like one of us." Another says, "Maybe he could help us fend off that wolf they found in California." Consciously or not, Meyer is tapping into old Hetch Hetchy themes. The addle-headed conservationists of the old *San Francisco Call* and *San Francisco Chronicle* cartoons are here depicted as sheep. Under the hooves of the sheep, the dam is green and pasture-like—here, finally, are those grasses and vines that San Francisco promised would cover the dam.

If the verbal debate over Hetch Hetchy found its old channels, then the graphic debate did, too.

In 1987, when Interior Secretary Don Hodel proposed restoring Hetch Hetchy, he triggered, first, the sequence of editorial-page eruptions that rumbles on today, and second, a 1988 National Park Service study, "Alternatives for Restoration of Hetch Hetchy Valley Following Removal of the Dam and Reservoir."

The three authors of the study were employees of Yosemite National Park: Richard Riegelhuth, chief of the Resources Management Division for Yosemite; Steve Botti, a resource management specialist; and

Jeff Keay, a wildlife biologist. The team sketched out three possible approaches to restoration. Alternative 1 was recovery without management, Alternative 2 was recovery with moderate management, and Alternative 3 was recovery with intensive management.

The Yosemite men began with a set of working assumptions: That sediments would be thin over the valley floor and would not need to be removed. That the river would reoccupy its original channel without human assistance. That Yosemite National Park would not attempt to insure the restoration of all plant taxa that originally grew in Hetch Hetchy. That the park would acquiesce to the invasion of many common non-native plants. That intensive herbicide spraying would not be permitted. That all animal species native to Hetch Hetchy, except grizzlies and wolves, now exist in viable populations immediately adjacent to the river canyon, and should volunteer to resettle the place. That some wildlife species may reoccupy recovering sites in such abundance as to inhibit vegetative recovery, or, if these species are predators, to inhibit recovery of other animal species. That earthworms and other soil organisms, as well as airborne invertebrates, will be naturally returned to Hetch Hetchy in a reasonable period of time without human intervention. That the aquatic ecosystem of the Tuolumne River will return to near pristine conditions without the help of humans. That the "bathtub ring" on the cliffs surrounding the reservoir will disappear naturally, in time, without management intervention.

The authors never say so explicitly, but their assumptions point with little deviation—only the slightest quivering, like a compass needle as it settles on north—toward Alternative 1, recovery without management. Hetch Hetchy, they suggest, is perfectly capable of restoring itself.

A good deal of subsequent thinking and writing has leaned sharply the other way, toward Alternatives 2 and 3, capital-intensive, labor-intensive, hands-on management by humans. In 2004, as an example, the University of Wisconsin published "Hetch Hetchy Valley: A Plan for Adaptive Restoration," a proposal amounting to Alternative 2.5, or thereabouts. The six Wisconsin authors describe adaptive restoration as "the design and implementation of a restoration project as a series of experiments, using knowledge from early experiments to revise subsequent experiments and improve subsequent restoration efforts." They point out that incremental drawdown of the reservoir would lend itself to incremental experimentation;

as the valley emerged slowly from the water, the restorers would adapt lessons learned from the successes and failures of earlier restoration experiments.

When it comes to the viability of Alternative 1, the "no-action alternative," as they call it, the Wisconsinites are not enthusiastic. "Because of the presence of invasive plant species in Yosemite, it is unlikely that the Hetch Hetchy Valley would be populated with the appropriate native species without an active restoration plan," they argue. They point out that invasive plants are often extremely good competitors in disturbed sites. "For this reason, we also rejected a partial-site restoration plan, where only some sites would be actively restored, while others were allowed to self-generate, and opted to use at least minimal restoration practices over the whole exposed area."

Nature, in other words, cannot be wholly trusted with any part of the restoration of Hetch Hetchy.

My own preference is for Alternative 1 and laissez-faire restoration. Ecosystems are resilient. The history of human attempts to manage ecosystems, despite the best of intentions, is a nearly unbroken saga of hubris, misapprehension, slapstick, unintended consequence, disaster, and loss.

Perhaps just one example will do: The Age of Dams was also the Age of Smokey the Bear. For more than half of the twentieth century, even as we were damming and channelizing our rivers, we were managing our forests according to principles of rigid fire suppression. "Only *you* can prevent forest fires," Smokey admonished us somberly from roadside signs in all our national woodlands. Preventing forest fires, it turns out, is a very bad idea. Many of our forests are naturally designed to burn. The buildup of understory fuel during the long suppression campaign has been catastrophic, as minor fires become conflagrations. This is now understood. For the past few decades, the Forest Service has been pursuing the antithetical strategy, with such determination that in season, as often as not, a pall of smoke from prescribed burns shadows the land. Best forest practice, as we now imagine it, has swung wildly back the other way.

Nature does not manage her systems like this. Natural regulation is not helter-skelter, with sudden changes in philosophy and reversals of protocol, full speed in one direction and then, whoops, full speed in the opposite. Nature manages smarter. Nature manages according to principles that began formulating at the beginning of time, or before, principles that will never be more than fragmentarily revealed to us.

My colleague Laura Cunningham, in researching her paintings of a future Hetch Hetchy for this book, has come to favor what amounts to Alternative 1.2. She believes that replanting black oaks might be a good idea, as otherwise they would take so long to reestablish mature groves. A look at the Merced River Canyon below Yosemite Valley—Hetch Hetchy's elevation exactly—has persuaded Cunningham that it will be impossible to suppress introduced Mediterranean grasses like wild oats and soft chess. This puts her in league with the Yosemite team of Riegelhuth, Botti, and Keay, who are antiherbicide, and in opposition to the Wisconsin team, in whose report one can find herbicides in the fine print.

"Trying to recreate the exact vegetation communities from mapping and historic photos may not be an ideal," Cunningham says, "since the valley is a new and different landscape after being flooded for so long. It might be better to let the ecologic and geophysical processes come into play, creating a more natural series of pioneering successional vegetation types that would find a balance eventually. I would anticipate a lot of wasted effort and expense if too much planning and planting were undertaken to try to recreate an exact time of the past. This would only be trying to recreate a snapshot in time."

This puts her in agreement with the ghost of my father. "It will probably be necessary to expedite the growth of grass for the first year," he wrote. He conceded that black oaks might have to be replanted. "Leave the rest to nature, and enjoy the spectacle of recovery. The jay and squirrel are experts at planting oaks. The wind can find the seeds that know how to grow wings. The wind also carries a whole inventory of spores, so there come the ferns, mosses, and lichens. Pines and other conifers know how to roll seeds downhill, and Hetch Hetchy Valley owes its existence to hills. Happily, Hetch Hetchy Valley is narrow, and the forces of renewal can creep across it."

This puts the two of them in the camp of John Muir.

"He is a man entirely without social sense," Congressman Kent complained of Muir a century ago. "With him, it is me and God and the rock where God put it, and that is the end of the story."

The rock where God put it is not a bad definition of wilderness.

Wilderness is that ever-diminishing realm where humans do not yet dominate and natural law applies. For the past century, our species has used Hetch Hetchy brutally—first logging the valley down to stumps and then flooding it wall to wall. Unnatural law has prevailed. The new century, as it is shaping up, offers the opportunity for something else here: to gently urge the hand of Man to release its iron grip. No more manipulation or intervention, not even with the best of intentions. It's an idea, anyway: let's see how the Sierra Nevada goes about resurrecting its own valley. That is the only experiment in adaptive restoration of any real interest to people of the Muir persuasion.

When someone of this faith looks out at the black oaks, Douglas fir, dogwoods, and incense cedars of a resurrected Hetch Hetchy Valley, that zealot does not want to see them growing where some restoration specialist has decided they should grow. We come here for something else. We want something more than can be found in Central Park or any other artificial landscape. We need to find those oaks and cedars and tall-grass meadows growing exactly where the Tuolumne River Canyon has decided they should grow. If we are painters or photographers or poets or religious people, we need it that way for spiritual reasons. If we are scientists, we need it that way as a laboratory for the study of natural law before unnatural practice garbled it. "Wilderness," as Nancy Newhall wrote, "holds answers to questions we have not yet learned how to ask."

It is obvious to people of the Muir persuasion—and I believe I speak for all of us—that if, or when, O'Shaughnessy Dam is decommissioned and Hetch Hetchy emerges again, the valley should come back as wilderness. There are more than a few of us who feel this way, especially when you multiply the present crop of us by all the generations of the future. And we are not all human beings; most of us are hawks and bears and deer and coyotes and weasels and water beetles.

AFTER THE DELUGE 4

The Western yellow or ponderosa pine, *Pinus ponderosa*—the tall tree for which, if you believe Chief Tenaya, Hetch Hetchy was named—grows best on flat or rolling land. In the Sierra Nevada it flourishes on the floors of U-shaped, glacier-carved valleys, just the sort of landscape that was flooded by O'Shaughnessy Dam. If Hetch Hetchy is drained, and if its reclamation proceeds more or less naturally, then the ponderosa-pine restoration will be led by chipmunks. Ponderosa-pine seedlings that spring up densely from forgotten chipmunk caches survive better, for some reason, than do solitary seeds sown by the wind. The Sierra Nevada is fortunate in having the greatest diversity of chipmunks on Earth.

For the seedling, whether dispersed by chipmunk or breeze, the first year is brutal. Frost heave, summer heat, and the browsing of mule deer thin the ranks mercilessly. The first mission of any newborn yellow pine, then, is to get purchase. The seedling sends down seven to twelve inches of taproot its first year, while sending up just two or three inches of tree. At eight years old, the mighty ponderosa pine towers up about sixteen inches. Into its early teens, it continues to grow slowly. In the reborn landscape of the Hetch Hetchy floor, at this early stage of the return of the yellow pine, the tree will call little attention to itself. Then, in its midteens, the tree shoots up, and for the next seventy-five to one hundred years grows fast, sometimes as much as two feet a year. When the ponderosa pine is fiftyish, it begins to bear cones profusely, offering up its winged seeds to chipmunks and the wind. It may never make the 220 feet that John Muir measured in one Sierra specimen, but it is headed that way.

"Of all Pines," wrote Muir, "this one gives forth the finest music to the winds." And of all pines, this one is also the most refulgent, its long reflective needles scintillant in

Unless otherwise noted, images throughout this chapter are by John Muir Laws from *The Laws Field Guide to the Sierra Nevada* and are used courtesy of Heyday.

sunshine, like upraised sabers, so that on a bright and windy day each tree makes a kind of cavalry charge. A century after the demise of the dam, the music and shimmer of big yellow pines will have returned to Hetch Hetchy.

The incense cedar, its sprays of foliage fashioned into a pillow by Muir, put him to sleep with its "spicy breath" on the eve of his first descent into Hetch Hetchy. It, too, will reseed itself, perhaps from Muir's very tree, and surely from other incense cedars growing now on the shores of the reservoir. The gray pine will do the same. For the Miwoks and Paiutes who hunted and foraged this valley, the seeds of the gray pine were nearly as important as Hetch Hetchy's acorns. The Natives ate the inner bark and young buds in springtime, cut out and consumed the soft cores of the cones when these were small and green, and ate the pine's resinous exudations like candy. The gray pine has fallen out of favor with humans as food, an unfortunate narrowing of the diet of our species, but a good thing for restoration, as there will be diminished competition for the ten-inch, sharp-spined cones that squirrels disassemble and plant across Hetch Hetchy.

The giant cones of the sugar pine, sometimes more than two feet long, will ripen in late summer and fall. The scales will open to send brown-winged seeds sailing out across the valley floor. The Douglas fir will return, as will the white fir, the cottonwood, the willow, the bigleaf maple, the Western azalea, the interior live oak, and the buckeye. At least a few black oaks—Kellogg's oak, as Muir called the tree—will come back, and with moderate restoration management, many black oaks. This big, black-trunked oak will leaf out pink for several weeks in spring, and in fall it will turn yellow-gold, providing most of what Hetch Hetchy offers in the way of hot color. The dogwood in springtime will open its four to six big, showy, petal-shaped bracts, green at first, then turning white. These pale, six-inch "flowers" will glow ghostly in the dimness of the conifers, like some sort of invasive orchid from the tropics.

Under the minimalist management strategy of Alternative 1, according to the predictions of the Yosemite Park team of Riegelhuth, Botti, and Keay, within two years of the draining of Hetch Hetchy Reservoir, extensive areas of the valley floor will be covered by grasses, sedges, and rushes,

Sugar pine cone

most of them non-native, growing most thickly on the sloping sides of the valley, as those slopes are drier and closer to seed sources. Dense nurseries of conifer and broadleaf seedlings will have sprung up in moister places at the edges of the valley.

Within ten years, some of the conifers established in the early succession will have grown to twenty feet tall. Stumps from the clear-cutting by the dam-builders will still be visible, but the general pattern of the vegetative communities to come will be evident. Areas that once were meadows will be reappearing as meadows again. Native plants—grasses, sedges, and rushes—will predominate in the meadows, excluding almost all non-native plants in the wettest areas. Non-natives will predominate elsewhere.

In fifty years, forest cover will be even more extensive than before the inundation. Some of the conifers established early will now be ninety feet tall. Pines and incense cedars more recently established will be rising in dense and expanding thickets, some of which will be invading the meadows.

In one hundred years, some of the ponderosa pines and incense cedars will stand 125 feet high and 5 feet in diameter at the base. Native plants will have reoccupied all their old favorable habitats, and all those species present in the old days will be here again. Non-native plants will have decreased in abundance.

Mammals will return, all the usual suspects mentioned in the resurrection scenarios for the valley—black bear, coyote, mule deer, mountain lion, squirrels—but also that large fauna of less familiar Sierra creatures that do not immediately come to mind.

The ringtail inhabits this zone and it will come in unseen. A small, nocturnal member of the family of the raccoons and coatis, the ringtail weighs two pounds or so, and has a bushy, banded tail longer than its body and huge eyes. It looks like a creature from somewhere else, Madagascar most likely, as if some Dr. Moreau of that island managed to cross a raccoon with a lemur. In gold rush days, in the foothills below Hetch Hetchy, ringtails were called "civet cats" by the miners, who domesticated them to rid their cabins of mice. Ringtails are cryptic creatures and belong to the night. In my own lifetime I have seen only one, an early-morning Sierra Nevada roadkill.

Ringtail tracks

The Mustelidae, the family of the weasels, will come to Hetch Hetchy. Curious, always testing the world around them, the mustelids are sure to investigate the valley floor as it emerges, all of them fierce and indomitable, most of them built sinuous and close to the ground, with hindquarters slightly elevated. A few family members slightly divergent from the basic weasel body plan—the striped skunk, spotted skunk, and badger—will find their way here. All the more weasel-like weasels will arrive, too, in their smoothly graduated succession of sizes: The short-tailed weasel, or ermine, a dwarf predator smaller than some of the larger chipmunks, at just two ounces. The long-tailed weasel, sometimes called the mountain weasel, the males weighing ten ounces, the females five. (In the cold months, both these small weasels, short-tailed and long, will bound across Hetch Hetchy snow in their white winter coats, invisible except for the black tips of their tails.) Then the two-pound pine marten, which will pursue and overtake squirrels in spiral, skittery, scratchy chases around the trunks of trees and out the branches. Then the ten-pound fisher, a scaled-up pine marten, heavy but acrobatic, fast both on the ground and in the trees, capable of overtaking and killing both squirrels and pine martens high above the earth. And finally, with any luck, the wolverine, the superweasel, king of the family Mustelidae.

As soon as the Tuolumne quickens and starts to flow freely through Hetch Hetchy again, the valley's aquatic ecosystem will begin reassembling itself. The current will recruit organisms upstream and transport them down, even as determined downstream volunteers work their way up against the flow.

The large water boatman and the small water boatman will swim in along the bottom, right side up, hind legs modified into oars and rowing a kind of breaststroke. Kirby's backswimmer and the small backswimmer will swim in at the surface, upside down, rowing a sort of backstroke. (At first glance, the boatmen and backswimmers will be hard to distinguish, except for this inversion.) The common water strider and the small water strider will glide in four-oared, either species built like a two-man racing shell. Standing spidery and weightless above the water, the striders will be visible less in themselves than by the dimples they make on the surface.

SPECIES IN HETCH HETCHY VALLE

Sketches by Laura Cunningham

ONE YEAR AFTER THE DAM

Larva of giant caddisfly

Larva of caddisfly *Desmona*

Larva of log-cabin caddisfly

Larva of caddisfly *Yphria*

The water scorpion will drift in, its head downward, its mantis-like raptorial forelimbs dangling. Connected to the surface by its snorkel—a long tube appended where the stinging tail would be in a genuine, arachnid sort of scorpion—it will scan the bottom for the small worms, isopods, and nymphs that are its prey. *Nymph,* the Greek word for stream sprite or spirit, is applied by science to the water-dwelling larvae of dragonflies, damselflies, stoneflies, and mayflies. This recycled Greek is apt when it comes to habitat, but all wrong when it comes to appearance. The dragonfly larva in particular—gigantic compound eyes, wicked biting mouthparts, spiky forked tail, gills in the anus—is more ogre than nymph. When it finally crawls up on its stem, splits its old skin, and emerges as a dragonfly—as fast as Tinker Bell, more maneuverable than any aircraft, its sides flashing iridescent red or blue or green in the mountain sunlight—then it might pass for a sprite, but not in its previous incarnation under the surface.

The larva of the giant caddisfly, upon arriving, will quarry the stream for the large sand grains with which it builds its protective case, and it will drag that little house over river-rounded stones exposed again by the flushing away of reservoir sediment.

The smaller larva of the caddisfly *Desmona* will build its case of smaller sand grains. On warm summer nights, when it reaches its fifth instar, *Desmona* will drag its house up on land—a concretion of stream-bottom materials come ashore—and it will graze on semiaquatic plants until the cold Sierra night drives it back into the water again.

The filter-feeding larva of the log-cabin caddisfly, *Brachycentrus,* will scavenge chips of Hetch Hetchy driftwood, square these off neatly with its jaws, and silk-cement them into the square, laminated case of its log cabin.

The predaceous larva of the caddisfly *Yphria*, perhaps the most primitive genus among these casemakers, will build its case of mica chips, slipshod and disorderly, like the log cabin of its cousin *Brachycentrus* after an earthquake.

The larva of the northern snail-shell caddisfly, *Helicopsyche* ("spiral-soul"), will coil its case of tiny sand grains into an uncanny facsimile of a snail shell—convergent evolution with the mollusks.

Larva of northern snail-shell caddisfly

The larva of the long-horned casemaker caddisfly, *Heteroplectron*, will take up residence in a hollow stick and graze on Tuolumne leaf litter.

The various larvae of the many caddisfly species in the genus *Limnephilus* ("Lake-lover") will build their cases in a wide assortment of styles, and from a multitude of materials: grass, or reeds, or moss, or woodchips, or sand, or the delicate and translucent shells of freshwater snails, or even the cast-off cases of smaller caddisfly larvae.

Larva of long-horned casemaker caddisfly

The free-living larva of the caddisfly *Rhycacophila* will hunt the bottom nude, with no protective case at all.

The Tuolumne, in finding its old channel again, will rediscover all those meanders where formerly it ran slowly, "often with a lingering expression, as if half inclined to become a lake," as Muir described it. In this peaceful water the predatory larva of the funnel-web caddisfly will weave its wide-mouthed, five-inch-long sack and wait inside for victims. The larva of the fingernet caddisfly will set out its tubular nets of fine silk on the undersides of rocks and feed on the small organic particles trapped in the mesh. In the slowest water, the larva of the soldier fly, armored in calcium carbonate, will scavenge and eat almost anything it can find.

Larva of caddisfly *Rhycacophila*

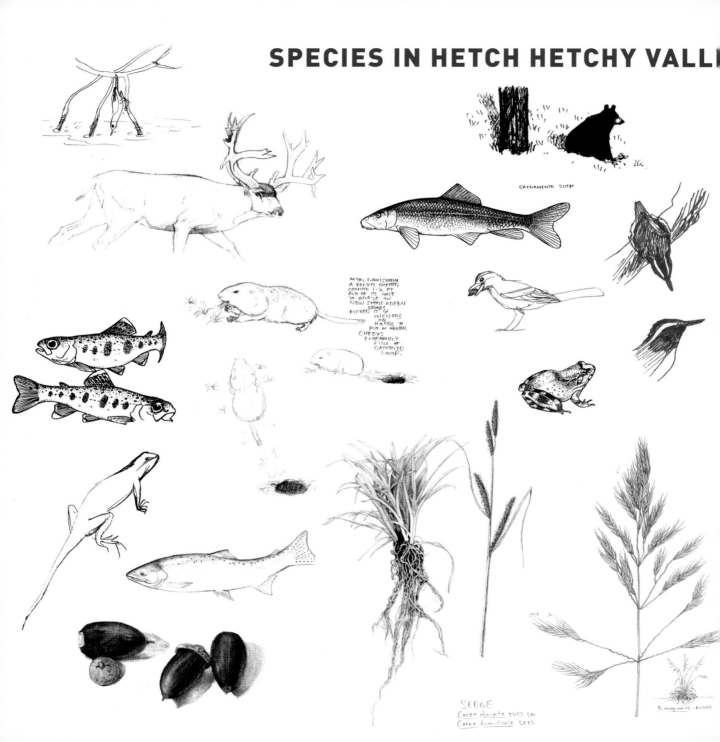

TWENTY YEARS AFTER THE DAM

ooh – A PRAIRIE FALCON FLEW BY ALONG RIDGE, LOW, → S. FLAPPING, GOING FAST.

LESSER GOLDFINCH ♂ PICKING SEEDS OF GREEN PHALARIS GRASS SPIKE

RED-TAIL HAWK

BARN SWALLOW

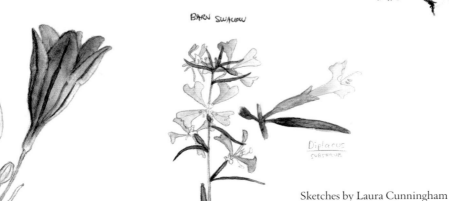

Diplacus (subshrub)

Clarkia RIBBONFLOWER

Sketches by Laura Cunningham

Pacific tree frog

As the reservoir subsides, the bar of glacier-polished granite that once divided the meadow of lower Hetch Hetchy from the forest of upper Hetch Hetchy will rise again, and the rapids that once ran over the bar will pick up speed and dash once more. The larva of the web-spinning caddisfly will cast its web into the fast water. The larva of the riffle beetle will forage on wood debris carried down by the riffles. The predaceous larvae of the alderfly, stonefly, and mayfly will hunt along the bottom, holding on against the current with the hooks on their feet. The larva of the black fly will filter-feed from river-rounded stones, belayed against the current by a silken safety line—convergent evolution with the human rock climber. In the fastest water, the tiny larvae of Comstock's net-winged midge will stick to the rock by its six suction cups, one each on the undersides of all six segments.

As the Tuolumne retraces its old course down the middle of the newly risen valley floor, and as Falls Creek, Tiltil Creek, Rancheria Creek, and the other tributaries cross that floor to join the river, riparian vegetation will reseed itself along the stream courses: willows, alders, dogwood. As this stream-loving flora takes hold, the Sierra's streamside fauna will follow. Yellow, yellow-rumped, orange-crowned, and black-throated warblers will linger in the riparian forests on their migration through the range. Black-headed grosbeaks and Bullock's orioles are likely to breed there. The mink should appear, and maybe the river otter, the two members of the weasel family most dependent on water. Amphibians will come: the California toad, the gregarious slender salamander, the California slender salamander, the Sierra newt, the arboreal salamander. The Pacific tree frog will surely show up, a tiny frog with a big voice, its choruses filling the Hetch Hetchy night.

The shrews will arrive. Trowbridge's shrew or the Yosemite shrew, *Sorex trowbridgii,* inhabits moist ground, meadows, and streamside vegetation at these elevations. The vagrant or wandering shrew, *Sorex vagrans*, is even more closely associated with water, and it lives in this zone as well. The water shrew, *Sorex palustris navigator* ("marsh navigator shrew") is the most aquatic of all, regularly diving from Sierra streamside vegetation to hunt the stream itself.

The shrews are the smallest of mammals, yet the most ferocious. They are not rodents, like mice, which generally dwarf them, but belong with moles in the primitive order Insectivora. More than anything they resemble small moles that have abandoned their burrows: the same velvety pelts, the same long, flexible snouts, the same tiny weak eyes, the same semiscrotal testes, the same frantic metabolism. As a general rule, a shrew consumes its own weight each day. Shrews in captivity have proved capable of eating their own weight every three hours. Shrews tear apart and gobble insects, spiders, centipedes, small reptiles and amphibians, snails, slugs, carrion, mice, and other shrews, feeding on all these things with a crazy insatiable desperation. Shrews are red in tooth and nail. The shrews of North America, indeed, belong in the group called red-toothed shrews—not red from blood, though that would make sense, but reddish from iron deposits reinforcing the enamel at the tips of the teeth, where the shrew's life of ceaseless predation causes the heaviest wear. Deprived of food for a few hours, shrews die of starvation. Some species dig tunnels, but many are too busy eating for that; instead they travel the burrows and underground galleries of moles, mice, and gophers, and they dash along the runways that mice and voles make in the grass. In these tunnels and runways, on encountering the rightful owners, shrews attack without hesitation rodents several times their own size. The saliva of some North American shrews is venomous. Enough neurotoxin can be extracted from the poison gland of a single shrew to kill two hundred mice.

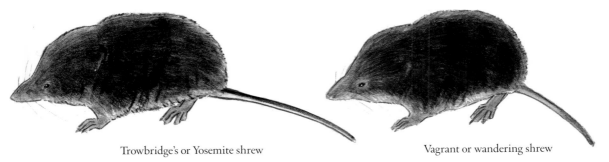

Trowbridge's or Yosemite shrew

Vagrant or wandering shrew

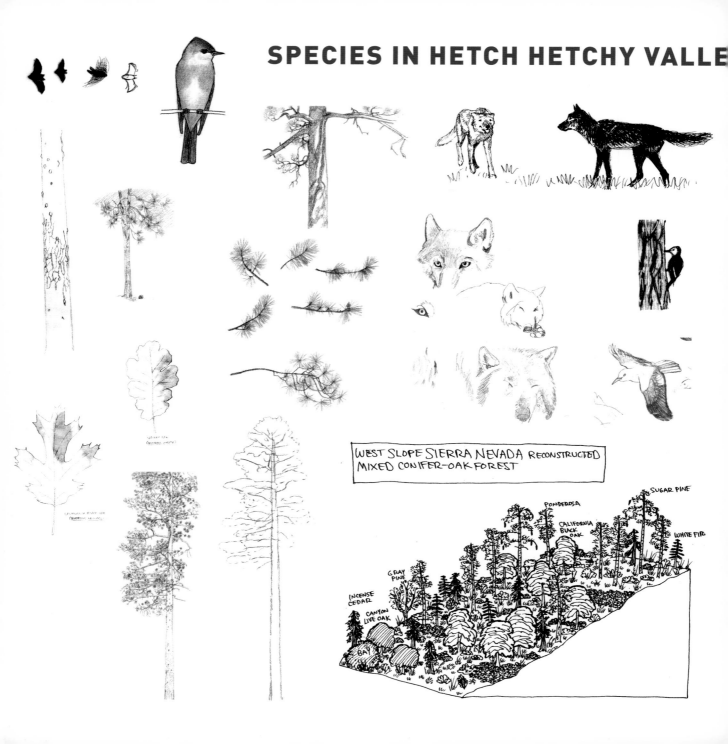

ONE HUNDRED YEARS AFTER THE DAM

WESTERN TANAGER ♀ FLYING

COOPER'S HAWK - BLUFF ATTACK ON UNPERTURBED SQUIRREL

BAND-TAILED PIGEONS OVER WOODED CANYON

FRONT 5" WIDTH

RED FOX GALLAVANTING ABOUT AFTER MICE/VOLES

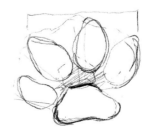

Sketches by Laura Cunningham

Shrews hunt nearly blind, depending almost entirely on their acute senses of feel, smell, and hearing. Some species of shrew echolocate, like dolphins and bats, emitting high-pitched squeaks and reading the rebounding signals to find their way and their prey.

Everything is fast with shrews, including reproduction. Copulation lasts ten seconds, gestation three weeks, childhood another three weeks or so, and then the young disperse to begin it all again. Females, sexually mature at six weeks, can bear as many as ten litters a year, becoming pregnant again within a day or two of giving birth and weaning one litter as the next is born. In the tropics shrews breed all year. In the temperate Sierra Nevada they allow themselves a winter break. Shrews do not hibernate. Evolution has discovered no way to turn down the thermostat of shrew metabolism, no means of reducing the average shrew's heart rate of 750 beats per minute *at rest* to the five heartbeats per minute of the hibernating chipmunk or the ten beats of the hibernating bear. Many shrew species respond to winter by shrinking. Under Sierra snows, they lose as much as half their body weight—a reduction not just of fat and muscle, but also of skull, bones, and internal organs. These abbreviated winter shrews, at half their summer size, must eat half again as much food to survive the cold season. For all their boldness and ferocity, shrews can be fatally jumpy and jittery. In their perpetual tachycardia—as many as 1,200 heartbeats a minute, under stress—they are vulnerable to sudden loud noises, and even the rumble of Sierra thunder can kill them. Life is measured in heartbeats and a shrew's life is never long. A two-year-old shrew is an ancient, its iron-tipped teeth worn down to nubs.

It is wearying just to think about it, the frantic pace of shrews.

Water shrew

The resurrection of Hetch Hetchy Valley will be a succession of many small yet momentous events. The arrival of *Sorex palustris navigator*, the water shrew, will be one such occasion, a great day in the reanimation of the Tuolumne River through this stretch of its canyon.

The water shrew is the largest of Sierra shrews, about the size of a house mouse. The feet of the water shrew are fringed with stiff hairs, a webbing especially noticeable on the hind feet, which are unusually large—swim paws, in effect. With each dive into the stream, tiny air bubbles trapped in the fur turn the shrew into an effervescent little silvery torpedo. It is as fast and agile underwater as a miniature otter, turning sharply and accelerating in pursuit of aquatic insects, small fish, and tadpoles. Often it leaves off swimming to scramble along the bottom, fighting its own buoyancy, poking its long snout under stones and sunken logs for whatever it can find there. Outside, its thick pelt keeps it warm in the icy, fast-running, snowmelt-fed Sierra streams. Inside it is warmed by the feverish shrew metabolism, its heart beating three hundred times per minute. When startled, it dives into the rapids, or, in stretches where the stream is smooth and glassy, it sprints away Jesus-like over the water, its webbed feet supporting it high and dry.

The Greeks had a name, *daimon,* for the nature spirits that they believed to reside in and watch over every hill, forest, spring, and stream in Greece. The Greeks did not invent the concept. All peoples once saw the world in this way; it was simply human to sense these invisible guardian presences everywhere in the landscape. The water shrew, at home in every part of the stream—scampering the bottom, swimming the midwater, walking the surface—is one candidate for daimon of the reborn Tuolumne through Hetch Hetchy.

But there is another candidate, a feathered analogue to the water shrew, a bird with a loud, clear, flutelike voice and a similar relationship to Sierra Nevada streams. No creature in the Sierra is more beloved, or more symbolic of fast water, than the dipper or water ouzel. A stocky, slate-gray, short-tailed songbird related to the wrens and thrushes, the water ouzel has adapted to life in the torrents. Its chunky shape gives it a high volume-to-surface ratio for retaining warmth. Its plumage is much denser than in other wrens. The oil gland, which all birds possess at the base of the tail, and which all birds, in preening, strop with their bills to lubricate and waterproof their feathers, is ten times larger in the ouzel than in other wrens and thrushes. The ouzel has scaly moveable hatch covers that seal its nostrils while underwater. It has an inner eyelid—a nictitating or "winking" membrane—which flashes white across the eye when it emerges. It has large, strong feet for gripping the stream cobbles as it walks the bottom against the current, hunting nymphs, larvae, and minnows.

Water ouzel or American dipper

The water ouzel seldom leaves the vicinity of water. In flying patrols above its streams, it does not shortcut over land from one branch to the next. This might save time, but it would waste opportunity. Instead the ouzel flies down one fork to the confluence, then up the other fork, keeping white, turbulent water always underneath. Its favorite nesting spot is behind a waterfall, the sort of secret headquarters that a Jules Verne or Tom Sawyer would favor. Like all water birds, it spends an inordinate amount of time preening and waterproofing. It can't seem to get enough of the stream. Often on finishing a dive it feels the need to water-bathe at the surface, shaking out its plumage, dunking its wings, scattering sunlit droplets of spray. On reaching water shallow enough to stand, or in flying to a streamside stone, the bird commences the odd, bobbing, genuflecting dance that earned it the name *dipper*. With each bob downward of the rounded breast, the tail flexes down, too, and often the nictitating membrane flashes white across the eye. When you first notice this white, sightless instant in the eye of a dipper, it is unsettling and creepy, as if for that instant the bird were possessed by a demon, but soon it seems normal. The dipping of the dipper is arrhythmic. The bird seems to be dancing to some syncopated beat it hears in the rush of the stream.

"The water ouzel, in his rocky home amid foaming waters, seldom sees a gun, and of all the singers I like him the best," John Muir wrote. "He is a plainly dressed little bird, about the size of a robin, with short, crisp, but rather broad wings, and a tail of moderate length, slanted up giving him, with his nodding, bobbing manners, a wrennish look. He is usually seen fluttering about in the spray of falls and the rapid cascading portions of the main branches of the rivers….

"How romantic and beautiful is the life of this brave little singer on the wild mountain streams, building his round bossy nest of moss by the side of a rapid or fall, where it is sprinkled and kept fresh and green by the spray! No wonder he sings well, since all the air about him is music; every breath he draws is part of a song, and he gets his first music lessons before he is born; for the eggs vibrate in time with the tones of the waterfalls. Bird and stream are inseparable."

Once the dam is gone, bird and stream will come together again. The first dipper will fly in over the Tuolumne as soon as the river rediscovers its gradient and begins to flow fast enough to please dippers.

The bird will wade the shallow margins, sticking its head under and periscoping for nymphs. In deeper pools, it will wing through the water like a penguin. Bursting in full flight from the torrent, it will alight on a streamside stone and commence its bobbing, tail-flexing dance: a grace note, a seal of approval, a benediction.

The next three spreads imagine Hetch Hetchy Valley one year, twenty years, and one hundred years after the removal of O'Shaughnessy Dam. Paintings by Laura Cunningham

Photograph by Barbara Brower

ABOUT THE AUTHOR

Kenneth Brower has written for *The Atlantic, Audubon, National Geographic, Canadian Geographic, The Paris Review, Reader's Digest, Smithsonian, Sierra, Islands,* and many other magazines. His emphasis has been environmental issues and the natural world. His work has taken him to all the continents. He is the author of *The Starship and the Canoe, Wake of the Whale, A Song for Satawal, Realms of the Sea, The Winemaker's Marsh, Freeing Keiko, The Wildness Within,* and many other books. He lives in Berkeley, California.

into California

About Heyday

Heyday is an independent, nonprofit publisher and unique cultural institution. We promote widespread awareness and celebration of California's many cultures, landscapes, and boundary-breaking ideas. Through our well-crafted books, public events, and innovative outreach programs we are building a vibrant community of readers, writers, and thinkers.

Thank You

It takes the collective effort of many to create a thriving literary culture. We are thankful to all the thoughtful people we have the privilege to engage with. Cheers to our writers, artists, editors, storytellers, designers, printers, bookstores, critics, cultural organizations, readers, and book lovers everywhere!

We are especially grateful for the generous funding we've received for our publications and programs during the past year from foundations and hundreds of individual donors. Major supporters include:

Anonymous (3); Acorn Naturalists; Alliance for California Traditional Arts; Arkay Foundation; Judy Avery; James J. Baechle; Paul Bancroft III; BayTree Fund; S. D. Bechtel, Jr. Foundation; Barbara Jean and Fred Berensmeier; Berkeley Civic Arts Program and Civic Arts Commission; Joan Berman; Buena Vista Rancheria/Jesse Flyingcloud Pope Foundation; John Briscoe; Lewis and Sheana Butler; California Civil Liberties Public Education Program; Cal Humanities; California Indian Heritage Center Foundation; California State Library; California State Parks Foundation; Keith Campbell Foundation; Candelaria Fund; John and Nancy Cassidy Family Foundation, through Silicon Valley Community Foundation; The Center for California Studies; Charles Edwin Chase; Graham Chisholm; The Christensen Fund; Jon Christensen; Community Futures Collective; Compton Foundation; Creative Work Fund; Lawrence Crooks; Nik Dehejia; Frances Dinkelspiel and Gary Wayne; The Durfee Foundation; Earth Island Institute; Eaton Kenyon Fund of the Sacramento Region Community Foundation; Euclid Fund at the East Bay Community Foundation; Foothill Resources, Ltd.; Furthur Foundation; The Fred Gellert Family Foundation; Fulfillco; The Wallace Alexander Gerbode Foundation; Nicola W. Gordon; Wanda Lee Graves and Stephen Duscha; David Guy; The Walter and Elise Haas Fund;

Coke and James Hallowell; Stephen Hearst; Historic Resources Group; Sandra and Charles Hobson; G. Scott Hong Charitable Trust; Donna Ewald Huggins; Humboldt Area Foundation; James Irvine Foundation; Claudia Jurmain; Kendeda Fund; Marty and Pamela Krasney; Guy Lampard and Suzanne Badenhoop; Christine Leefeldt, in celebration of Ernest Callenbach and Malcolm Margolin's friendship; LEF Foundation; Thomas Lockard; Thomas J. Long Foundation; Judith and Brad Lowry-Croul; Kermit Lynch Wine Merchant; Michael McCone; Nion McEvoy and Leslie Berriman; Michael Mitrani; Moore Family Foundation; Michael J. Moratto, in memory of Ernest L. Cassel; Richard Nagler; National Endowment for the Arts; National Wildlife Federation; Native Cultures Fund; The Nature Conservancy; Nightingale Family Foundation; Northern California Water Association; Pacific Legacy, Inc.; The David and Lucile Packard Foundation; Patagonia, Inc.; PhotoWings; Alan Rosenus; The San Francisco Foundation; San Manuel Band of Mission Indians; Greg Sarris; Savory Thymes; Sonoma Land Trust; Stone Soup Fresno; Roselyne Chroman Swig; Swinerton Family Fund; Thendara Foundation; Sedge Thomson and Sylvia Brownrigg; TomKat Charitable Trust; The Roger J. and Madeleine Traynor Foundation; Lisa Van Cleef and Mark Gunson; Patricia Wakida; Whole Systems Foundation; Wild by Nature, Inc.; John Wiley & Sons, Inc.; Peter Booth Wiley and Valerie Barth; Bobby Winston; Dean Witter Foundation; The Work-in-Progress Fund of Tides Foundation; and Yocha Dehe Community Fund.

Board of Directors

Guy Lampard (Chairman), Richard D. Baum, Barbara Boucke, Steve Costa, Nik Dehejia, Peter Dunckel, Theresa Harlan, Marty Krasney, Katharine Livingston, Michael McCone (Chairman Emeritus), Sonia Torres, Michael Traynor, Lisa Van Cleef, and Patricia Wakida.

Getting Involved

To learn more about our publications, events, membership club, and other ways you can participate, please visit www.heydaybooks.com.